Introduction to the Principles and Practice of Soil Science

R. E. WHITE

University Lecturer in Soil Science and Fellow of
St Cross College, Oxford

BLACKWELL SCIENTIFIC PUBLICATIONS

OXFORD LONDON EDINBURGH MELBOURNE

© 1979 by Blackwell Scientific Publications
Osney Mead, Oxford, OX2 0EL
8 John Street, London, WC1N 2ES
9 Forrest Road, Edinburgh, EH1 2QH
P.O. Box 9, North Balwyn, Victoria, Australia

First published 1979

Typeset by Enset Ltd.
Midsomer Norton, Bath and
Printed and bound in Great Britain by
the Alden Press, Oxford

British Library Cataloging in Publication Data

White, R E
    Introduction to the principles and practice of soil science.
    1. Soils
    1. Title
    631.4   S591
    ISBN 0-632-00052-X

Contents

PART 1 · THE SOIL HABITAT

'The soil is teeming with life. It is a world of darkness, of caverns, tunnels and crevices, inhabited by a bizarre assortment of living creatures . . .'
J.A.Wallwork (1975) in
The Distribution and Diversity of Soil Fauna.

PART 2 · PROCESSES IN THE SOIL ENVIRONMENT

'Soils are the surface mineral and organic formations, always more or less coloured by humus, which constantly manifest themselves as a result of the combined activity of the following agencies; living and dead organisms (plants and animals), parent material, climate and relief.'
V.V.Dokuchaev (1879), quoted by
J.S.Joffe in *Pedology.*

PART 3 · UTILIZATION OF SOILS

'And earth is so surely the food of all plants that with the proper share of the elements, which each species requires, I do not find but that any common earth will nourish any plant.'
Jethro Tull (1733) in
Horse-hoeing Husbandry,
republished by William Cobbett in 1829

Preface

Soil science has a poor image. Its contribution to the welfare of modern society has been overlooked by the commentators who shape public opinion. The layman's knowledge of soil is derived from trivial articles in the gardening columns of newspapers and magazines, from which one might conclude that adding compost to build up soil humus is the panacea for all ills, from the scorching of plants during a hot, dry summer to the depredations of insects and disease. Little wonder that soil studies are jocularly relegated to the realm of 'muck and mystery'.

But research into the replenishment of soil water, the microstructure of clay domains or the biology of the rhizosphere can be intellectually as stimulating as deciphering the genetic code, or smashing atoms, and equally important to Man's well-being. The study of soil draws heavily on the established principles of the physical, biological and earth sciences, and I therefore make no apology for assuming that readers of this book have a knowledge of high school physics, chemistry and biology, together with modest mathematical skills.

Not only does soil science deserve recognition as a discipline in its own right, but it also makes a fundamental contribution to the science of agriculture and forestry, ecology and the environment, physical geography, civil engineering and archaeology. Teachers in each subject naturally emphasize certain aspects of soil more than others. Consequently it is difficult in one book to satisfy the special requirements of pedologists, ecologists and geographers, and to strike the right balance in the selection of examples of soil behaviou between agricultural and natural ecosystems.

In an attempt to solve these problems, I have divided the book into three parts. Part I introduces the soil in broad outline and describes its components in detail. Part II highlights the important physiochemical and biological processes which determine soil behaviour—it is essentially a story of the interaction of water, gases and solutes with the surfaces of finely divided mineral and organic material. Using this basic knowledge, Part III describes how the soil must be managed to provide the food and other products necessary for everyday life. Parts I and II should be of interest to any student, irrespective of his or her special subject, who wishes to know more about soil *per se*. Although Part III is more appropriate to the soil scientist and agronomist, I hope it will provide the ecologist with a better understanding of the agriculturalist's viewpoint.

In the end, I shall be satisfied if this book encourages its readers to learn more about the soil. By so doing, I hope they will recognize the need for soil, like other natural resources, to be managed in a way that provides for people today, while at the same time preserving a productive potential and amenity value for posterity.

R.E. White

Oxford, January 1979

vii

Acknowledgements

I am greatly indebted to many people who helped in the preparation of this book: to Philip Beckett without whose inspiration and constant encouragement I would neither have begun nor completed the task; and to my colleagues Peter Nye, Frank Thompson, Roger Hall, Bill Handley and others farther afield—David Rowell, Grant Thomas, Doug Lathwell, Duncan Greenwood and T. E. Tomlinson, who criticized various chapters. Jeni Hughes, H. F. Woodward, John Baker, John Habgood, Olive Godwin, Rosetta Plummer, Cathy McIntosh and my wife Annette gave invaluable help with the illustrations, literature research and typing the manuscript, and Robert Campbell of Blackwell Scientific Publications can justly claim to represent a humane publishing house. Finally, I appreciate the patience and forbearance that my family have shown during the writing of this book.

I should like to thank those scientists, mentioned individually beneath the appropriate figures, who either sent me photographs to be used as illustrations or gave permission for their originals to be reproduced. Acknowledgement is also due to the publishers who gave permission for original material to be reproduced from the following publications:

Figure 1.2—from *Journal of the Australian Institute of Agricultural Science* **42**, 75–93 (1976).

Figure 2.4—from *Methods for Analysis of Irrigated Soils* (Ed.) J. Loveday, Tech. Comm. No. 54, Commonwealth Agricultural Bureaux, Farnham Royal.

Figures 2.6, 2.7—from *Environmental Chemistry* by J.W. and E.A. Moore, Academic Press, London.

Figure 3.2—from *Soil Microbiology* (Ed.) N. Walker, Butterworth, London.

Figure 3.4—from *Nature New Biology* **246**, 269–271 (1973), Macmillan (Journals), London.

Figure 3.5—from *Soil Conditions and Plant Growth* by E.W. Russell, Longman, London; with permission of the Director, Rothamsted Research Station.

Figure 3.6—from *Biology of Earthworms* by C.A. Edwards and J.R. Lofty, Chapman and Hall, London.

Figure 4.8—from *Fabric and Mineral Analysis of Soils* by R. Brewer, Wiley and Sons, London and Krieger Publishing Co., New York.

Figure 5.3—from *Agriculture and Water Quality* MAFF Tech. Bull. No. 32, HMSO, London.

Figures 11.6, 11.10—from *Journal of Soil Science* **23**, 363–380 (1972), Oxford University Press, Oxford.

Figure 12.3—from *Chemistry and Technology of Fertilizers* by A.V. Slack, Wiley and Sons, London.

Figure 12.5—from *Soil Fertility and Fertilizers* by S.L. Tisdale and W.L. Nelson, Macmillan Publishing, New York.

Figure 12.8—from *Pesticides and Human Welfare* (Eds.) D.L. Gunn and J.G.R. Stevens, Oxford University Press, Oxford.

Units of Measurement and Abbreviations used in this Book

SI units

Basic unit	Symbol
metre	m
kilogram	kg
second	s
ampere	A
kelvin	K
mole	mol

Derived and SI related units

Unit	Symbol	Value
newton	N	$kg\ m\ s^{-2}$
joule	J	N m
pascal	Pa	$N\ m^{-2}$
volt	V	$J\ A^{-1}\ s^{-1}$
siemen	S	$A\ V^{-1}$
coulomb	C	A s
hectare	ha	$10^4\ m^2$
litre	1 or dm^3	$10^{-3}\ m^3$
tonne	t	$10^3\ kg$
bar	bar	$10^5\ Pa$

Non-SI units used in soil science

Physical quantity	Unit	Symbol	Value
length	Ångstrom	Å	$10^{-10}\ m$
temperature	degree Celsius	°C	$K-273$
concentration	moles per litre	M	$mol\ l^{-1}$
cation exchange capacity	milli-equivalent per 100 gram	me per 100 g	

radioactivity	Curie	Ci	3.7×10^{10} disintegrations s^{-1}
electrical conductivity	millimho per cm	mmho cm^{-1}	$10^{-3}\ S\ cm^{-1}$

Prefixes to SI and derived SI units

kilo-	k	10^3
deca-	da	10^1
deci-	d	10^{-1}
centi-	c	10^{-2}
milli-	m	10^{-3}
micro-	μ	10^{-6}
nano-	n	10^{-9}
pico-	p	10^{-12}

ABBREVIATIONS USED IN THIS BOOK

API	Air photo interpretation
an	annum (year)
AR	Activity ratio
AWC	Available water capacity
B.D.	Bulk density
B.P.	Before present
c.	about, approximately
CEC	cation exchange capacity
concn.	concentration
CSIRO	Commonwealth Scientific and Industrial Research Organization
D.M.	Dry matter
EC	Electrical conductivity

Abbreviations used in this book *continued*

E.I.	Soil erodibility index
ESP	Exchangeable sodium percentage
ET	Evapotranspiration
FAO	Food and Agriculture Organization
FC	Field capacity
hr	hour
IAEA	International Atomic Energy Agency
IEP	Isoelectric point
IHP	Inner Helmholtz plane
IR	Infiltration rate
K.E.	Kinetic energy
LHS	Left-hand-side
LR	Leaching requirement
MAFF	Ministry of Agriculture, Fisheries and Food
M.W.	Molecular (formula) weight
o.d.	oven-dry
OHP	Outer Helmholtz plane
ppm	parts per million
PSR	Pore space ratio
PZC	Point of zero charge
RHS	Right-hand-side
RQ	Respiratory quotient
SAR	Sodium adsorption ratio
S.G.	Specific gravity
SMD	Soil moisture deficit
soln	solution
sp, spp	species (singular; plural)
SSEW	Soil Survey of England and Wales
TDS	Total dissolved solids

U.K.	United Kingdom
UNESCO	United Nations Education, Scientific and Cultural Organization
U.S.A.	United States of America
USDA	United States Department of Agriculture
USLE	Universal soil loss equation
WP	Wilting point
yr	year

Some chemical and miscellaneous symbols

2,4-D	2,4-dichlorophenoxyacetic acid
DAP	Diammonium phosphate
DCPD	Dicalcium phosphate dihydrate
DDT	Dichlorodiphenyltrichloroethane
GRP	Ground rock phosphate
HA	Hydroxyapatite
KCP	Potassium calcium pyrophosphate
KMP	Potassium metaphosphate
MAP	Monoammonium phosphate
MCP	Monocalcium phosphate
OCP	Octacalcium phosphate
PVA	Polyvinyl alcohol

( )	denotes activities
$\simeq$	approximately equal to
$\sim$	of the order of
<	less than
>	greater than
$\leq$	less than or equal to
ln	$\log_e$

Part 1
The Soil Habitat

'The soil is teeming with life. It is a world of darkness, of caverns, tunnels and crevices, inhabited by a bizarre assortment of living creatures . . .'
J.A. Wallwork (1975) in
The Distribution and Diversity of Soil Fauna.

Chapter 1
Introduction to
the Soil

1.1 SOIL IN THE MAKING

With the exposure of rock to a new environment—by the extrusion and solidification of lava, uplift of sediments, the recession of water or the retreat of a glacier—a soil begins to form. Decomposition proceeds inexorably towards a decreased free energy and increased entropy. The *free energy* of a closed system, such as a rock fragment, is that portion of its total energy which is available for work, other than work done in expanding its volume. Part of the energy released in a spontaneous reaction, such as rock weathering, appears as entropic energy, and the entropy is a measure of the degree of disorder of the system. For example, as the rock weathers, crystalline minerals are broken down into simple molecules and ions, some of which are leached out or escape as gases.

Weathering is hastened by the appearance of primitive forms of plant life on the rock surfaces. These plants—lichens, mosses and liverworts—can store radiant energy from the sun as chemical energy in the products of photosynthesis. Lichens, which are a symbiotic association of an alga and fungus, are also able to 'fix' atmospheric nitrogen and incorporate it into plant protein. Thus, on the death of each generation of these primitive plants, some of the rock elements and a variety of complex organic molecules are returned to the weathering surface where they nourish the succession of organisms gradually colonizing the embryonic soil.

Figure 1.1. Stages in soil formation on a calcareous parent material in a humid temperate climate.

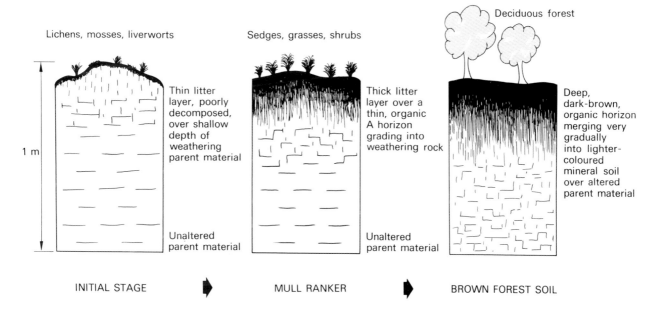

Lichens, mosses, liverworts

Sedges, grasses, shrubs

Deciduous forest

1 m

Thin litter layer, poorly decomposed, over shallow depth of weathering parent material

Unaltered parent material

Thick litter layer over a thin, organic A horizon grading into weathering rock

Unaltered parent material

Deep, dark-brown, organic horizon merging very gradually into lighter-coloured mineral soil over altered parent material

INITIAL STAGE ▶ MULL RANKER ▶ BROWN FOREST SOIL

A very simple example is that of soil formation under the extensive deciduous forests of the cool humid parts of Europe, Asia and North America, on calcareous deposits exposed by the retreat of the Pleistocene ice cap. The profile development is summarized in Figure 1.1. The initial state is little more than a thin layer of weathered material stabilized by primitive plants. Within one to two centuries, as the organo-mineral material accumulates, more advanced species of sedge and grass which are adapted to the harsh habitat appear, and the developing soil is described as a *mull ranker*. The dead plant remains foster the growth of pioneering micro-organisms and animals, which gradually increase in abundance and variety. The litter deposited on the surface is mixed into the soil by burrowing animals and insects, where its decomposition is hastened. The eventual appearance of larger plants— shrubs and trees, with their deeper rooting habit, pushes the zone of rock weathering farther below the soil surface. After perhaps a thousand years and more, a mature *brown forest soil* emerges. We shall return to the topic of soil formation, and the processes involved, in Chapters 5 and 9.

1.2 CONCEPTS OF SOIL

The soil is at the interface between the *atmosphere* and *lithosphere* (the mantle of rocks making up the Earth's crust). It also interfaces with bodies of fresh and salt water (collectively called the *hydrosphere*). The soil sustains the growth of many plants and animals, and so forms a part of the *biosphere*.

There is little merit in attempting a rigorous definition of soil, because of the complexity of its make-up and of the physical, chemical and biological forces to which it is exposed. Nor is it necessary to do so, for the soil means different things to different users. To the geologist and engineer, the soil is no more than finely-divided rock material. The hydrologist sees the soil as a storage reservoir affecting the water balance of the catchment he is studying, while the ecologist may be interested in only those soil properties which influence the growth and distribution of different plant species. The farmer is naturally concerned about the multi-

plicity of ways in which the soil can influence the growth of his crops and the health of his livestock, although frequently his interest will not extend below the 20 to 25 cm depth of soil which is disturbed by the plough.

In view of this wide spectrum of potential user need, it is appropriate in introducing the topic of soil to readers, perhaps for the first time, to review briefly the evolution of man's relationship with the soil and identify some of the past and present concepts of soil.

Soil as a medium for plant growth

Man's exploitation of the soil for food production began some two or three thousand years after the close of the last Pleistocene ice age, which occurred about 11,000 years B.P. (before present). Neolithic man and his primitive agriculture spread outwards from settlements in the fertile crescent embracing the ancient lands of Mesopotamia and Canaan (Figure 1.2) and reached as far as China and the Americas within a few thousand years. In China, for example, the earliest records of soil survey (4,000 yr B.P.) show how the fertility of the soil was used as a basis for levying taxes on the incumbent landholders. The study of the soil was a practical exercise of everyday life, and the knowledge of soil husbandry that had been acquired by Roman times was faithfully passed on by peasants and landlords, with little innovation, until the early 18th century.

From that time onwards, however, the rise in demand for agricultural products in Europe was dramatic. Conditions of comparative peace, and rising living standards as a result of the Industrial Revolution, further stimulated this demand throughout the 19th century. The period was also one of great discoveries in the sciences of physics and chemistry, the implications of which sometimes burst with shattering effect on the conservative world of agriculture. In 1840, von Liebig firmly debunked the 'humus theory' of plant nutrition, and established that plants absorbed nutrients as inorganic compounds from the soil. In the 1850s Way discovered the process of cation exchange in soils; and during the years from 1860 to 1890 eminent bacteriologists including Pasteur, Warington and Winogradsky elucidated the role of micro-organisms

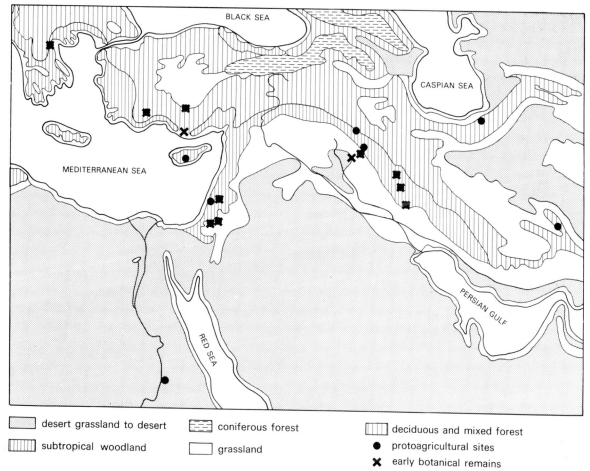

Figure 1.2. Sites of primitive settlements in the Middle East (after Gates, 1976).

in the decomposition of plant residues and the conversion of ammonia to nitrate.

Over the same period, botanists such as von Sachs and Knop, by careful experimentation in water culture, identified the major elements which were essential for healthy plant growth. Agricultural chemists drew up balance sheets of the quantities of these elements taken up by crops, and by inference, the quantities which needed to be returned to the soil in fertilizers or animal manure to sustain growth. This approach, whereby the soil is regarded as a relatively inert medium providing water, mineral* ions and physical support for plant growth, has been called the 'nutrient bin' concept.

Soil and the influence of geology

The pioneering chemists who investigated the soil's ability to supply nutrients to plants tended to see the soil merely as a medium in which chemical and biochemical reactions occurred. They little appreciated the soil as part of a landscape, moulded by natural forces acting on the Earth's crust. Great contributions were made to our knowledge of the soil *in situ* by the

* The term 'mineral' is used in two contexts: firstly, as an adjective referring to the inorganic constituents of the soil (ions, salts and particulate matter); secondly, as a noun referring to specific inorganic compounds found in rocks and soil, such as quartz and feldspars.

application of the field survey methods of the new science of geology in the late 19th century. Geologists defined the mantle of loose, weathered material on the Earth's surface as the *regolith*, of which only the upper few decimetres could be called soil, since this was the usual extent to which organic material was incorporated (Figure 1.3). Below the soil was the subsoil which was

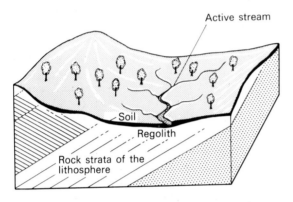

Figure 1.3. Soil development in relation to the landscape and underlying regolith.

largely devoid of organic matter. But the mineral matter of both soil and subsoil was recognized as being derived from the weathering of underlying rock. This led to intensive investigations of the relationship between various soils of the landscape and the geology of the underlying rocks, with the result that soils were loosely classified in geological terms, such as loessial, glacial, alluvial, granitic or marly (derived from limestone).

The influence of Russian soil science

Russia at the end of the 19th century was a vast country covering many climatic zones in which, on the whole, crop production was limited not by infertility of the soil, but by the primitive methods of agriculture. Russian soil science was therefore concerned not so much with soil fertility, but with observing soils in the field, and studying relationships between the properties of a soil and the environment in which it had formed. From 1870 onwards, Dokuchaev and his school clearly distinguished the soil from the loose material below, the *parent material*, by its characteristic physical,

chemical and biological make-up. This was the beginning of the science of *pedology.**

The impact of the Russian work on the development of pedology in the Western World was considerable. Scientists who had previously studied only small samples in the laboratory, or who were interested only in those soil properties that might influence plant growth, began to appreciate the multiplicity of the forces acting on the soil—the so-called *soil-forming factors*, which are discussed in Chapter 5. The idea arose that under any particular combination of soil-forming factors, a unique physiochemical and biological environment was established which led to the development of a distinctive soil body—the process of *pedogenesis*.

Fundamental features of the soil were identified and described by a new set of terms. The vertical dimension, exposed by excavation from the surface to the parent material, constitutes the *soil profile*. Layers in the soil distinguished on the basis of colour, hardness, the occurrence of included structures and other visible or tangible properties, are called *horizons*. The upper layer, from which materials are generally washed downwards, is described as *eluvial*; lower layers in which these materials can accumulate are called *illuvial*. In 1932, an international gathering of soil scientists adopted the notation of *A* and *B* for the eluvial and illuvial horizons, and *C* horizon for the parent material (Figure 1.4). The *A* and *B* horizons comprise the *solum*. If unweathered rock exists below the parent material it is labelled *bedrock R*. The organic litter on the surface, but not incorporated in the soil, is designated the *L* layer.

Soil genesis is now known to be much more complex than this early work would suggest. It is acknowledged, for example, that most soils are *polygenetic* in origin; that is, they have undergone successive phases of development due to changes in climate and other environmental factors with time. Nevertheless, the Russian approach was a considerable advance on traditional thinking, and the recognition of the intimate relationship between soil and the environment encouraged soil scientists in their efforts to survey and

---

* From the Greek word for ground or earth.

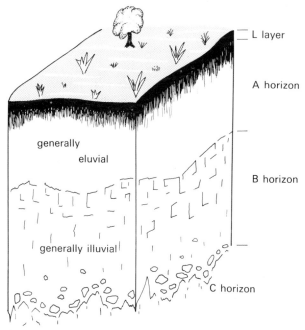

Figure 1.4. Diagram of a soil profile.

map the distribution of soils in many countries. The great range of soil morphology that was described was in turn a stimulus to studies of pedogenesis, an understanding of which, it was believed, would enable the mass of field data on soils to be collated in a systematic manner. Thus, Russian soil science provided the inspiration for many of the early attempts at soil classification.

A contemporary view of soil

Between the two World Wars of the 20th century, the philosophy of the soil as a 'nutrient bin' was much in evidence, particularly in the Western World. More and more land was brought into cultivation in areas marginal for crop production because of limitations of climate, soil and topography. With the balance between crop success and failure rendered even more precarious than in favoured agricultural areas, the age-old problems of soil loss by wind and water erosion, the encroachment of weeds and the accumulation of salts in irrigated lands became more serious and exacted a heavy toll of arable soils. In two decades or so after World War II, the escalating demands for food, fibre

and forest products by a world population (which at the time of writing has reached an estimated 4,000 million) led to increased use of fertilizers to improve crop yields, and pesticides to control pests and diseases. Such practices inevitably resulted in a steady rise of fertilizer residues (particularly nitrates) in surface and underground waters, as well as the widespread dispersal of the more stable pesticides throughout the biosphere, and their accumulation to potentially toxic levels in some species of birds and fish.

More recently, however, there has been a change of emphasis in the exploitation of world soil resources. The need for a compromise between optimum crop and animal production on the one hand, and conservation of a valuable resource and of man's environment on the other, has been acknowledged by scientists, planners and producers. Most soil scientists now recognize the soil as a natural body, clearly distinguished from inert rock material by

(1) the presence of plant and animal life;
(2) a structural organization which reflects the action of pedogenic processes;
(3) a capacity to respond to environmental change which may alter the balance between gains and losses within the profile, and predetermine the formation of a different kind of soil in equilibrium with the new set of environmental conditions.

Unlike other natural bodies, however, the soil has no fixed inheritance, but is dependent upon the conditions prevailing during its formation. Nor is it possible to define unambiguously the boundaries of the soil body—the soil atmosphere is continuous with the air above ground; many of the soil organisms live as well on the surface as within the soil; the litter layer usually merges gradually with the decomposed organic matter of the soil and likewise, the boundary between soil and parent material is difficult to demarcate. It is common, therefore, to speak of the soil as a *three dimensional body which is continuously variable both in time and space.*

1.3 COMPONENTS OF THE SOIL

As we have seen, a combination of physical, chemical and biotic forces acts on organic and weathered rock fragments to produce a soil with a porous fabric that can retain water and gases. The water contains dissolved organic and inorganic solutes and is called the *soil solution*. Whilst the soil air consists primarily of nitrogen and oxygen, it usually contains higher concentrations of carbon dioxide than the atmosphere, and traces of other gases which are by-products of microbial metabolism. The relative proportions of the four major components—*mineral matter*, *organic matter*, *water* and *air*—may vary widely, but generally lie within the ranges indicated in Figure 1.5. These

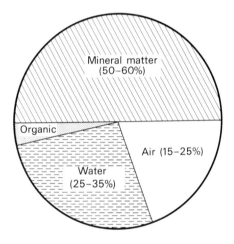

Figure 1.5. Proportions of the main soil components.

components are discussed in more detail in the subsequent chapters of Part 1.

1.4 SUMMARY

Soil forms at the interface between the atmosphere and the consolidated or loose deposits of the Earth's crust. Physical and chemical weathering, denudation and re-deposition combined with the activities of a succession of colonizing plants and animals moulds a distinctive soil body from the milieu of rock minerals in the parent material. The process of soil formation, called *pedogenesis*, culminates in the remarkably variable differentiation of the soil material into a series of horizons which constitute the *soil profile*. The horizons are distinguished, in the field at least, by their visible and tangible properties such as colour, hardness and structural organization. The intimate mixing of mineral and organic materials to form a porous fabric that is permeated by water and air creates a favourable habitat for a variety of plant and animal life.

FURTHER READING

EPSTEIN E. (1972) *Mineral Nutrition of Plants: Principles and Perspectives.* Wiley, New York.
KELLOGG C.E. (1950) Soil. *Scientific American* **821**, 2–11.
RUSSELL E.W. (1973) *Soil Conditions and Plant Growth.* 10th ed. Longmans, London.
SIMONSON R.W. (1968) Concepts of Soil. *Advances in Agronomy* **20**, 1–47.
UNITED STATES DEPARTMENT OF AGRICULTURE YEARBOOK (1938) *Soils and men.* U.S. Government Printing Office, Washington D.C.

Chapter 2
The Mineral Component
of the Soil

2.1 THE SIZE RANGE

Mineral fragments in soil vary enormously in size from boulders and stones down to sand grains and minute particles that are beyond the resolving power of an optical microscope (<0.2 μm in length or diameter). Particles smaller than c. 1 μm are classed as *colloidal*. If they do not settle quickly when mixed with water, they are said to form a colloidal solution or *sol*, or a *suspension* if they settle within a few hours. Colloidal solutions are distinguished from true solutions (dispersions of ions and molecules) by the *Tyndall effect*. This occurs when the path of a beam of light passing through the solution can be seen from either side, at right angles to the beam, indicating a scattering of the light rays.

An arbitrary division is made by size-grading soil into material which passes through a sieve with 2 mm diameter holes—the *fine earth*, and that which is retained on the sieve—the *stones* or *gravel*. Several categories of stone between 2 and 600 mm diameter are described: larger fragments are called boulders. The separation by sieving is carried out on air-dry soil which has been gently ground by mortar and pestle, or crushed between wooden rollers, to break up the obvious *aggregates* of smaller particles. Air-dry soil is soil allowed to dry in air at ordinary temperatures.

Particle-size distribution of the fine earth
The distribution of particle sizes determines the soil *texture* which may be assessed subjectively in the field, or more rigorously by particle-size analysis in the laboratory.

SIZE CLASSES
In any natural material, such as soil, one expects a continuous spectrum of particle sizes. This is called a frequency distribution, some examples of which are given in Figure 2.1. In practice, however, it is con-

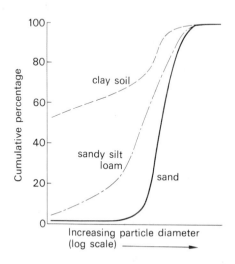

Figure 2.1. Frequency distributions of soil particle sizes in a clay, sandy silt loam and sandy soil.

venient to subdivide this continuous distribution into several class intervals which define the size limits of the *sand*, *silt* and *clay* fractions. The extent of this subdivision, and the class limits chosen, vary from country to country and between institutions within countries. The major systems in use today are those adopted by the Soil Survey Staff of the USDA, the British Standards Institution and the International Society of Soil Science. These are illustrated in Figure 2.2. All three systems set the upper limit for clay at 2 μm, but differ in the upper limit chosen for silt and the way in which the silt and sand fractions are subdivided.

FIELD TEXTURE
A soil surveyor determines texture by moistening a soil sample until it glistens and kneading it between fingers and thumb until the aggregates are broken down and the soil grains thoroughly wetted. The proportions of sand, silt and clay are estimated according to the following qualitative criteria:

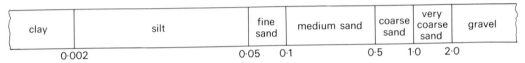

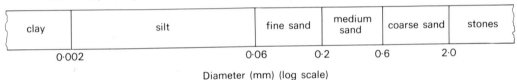

Diameter (mm) (log scale)

Figure 2.2. Particle-size classes most widely adopted internationally.

Coarse sand grains are large enough to grate against each other and can be detected individually by sight and feel.

Fine sand grains are much less obvious, but when more than about 10 per cent of the sample they can be detected by feel.

Silt grains cannot be detected by feel, but their presence makes the soil feel smooth and soapy, and only very slightly sticky.

Clay is characteristically sticky, although some dry clays require much moistening and kneading before they develop their maximum stickiness.

High organic matter contents tend to reduce the stickiness of clayey soils and to make sandy soils feel more silty. Finely divided calcium carbonate also gives a silt-like feeling to the soil.

The soil is assigned to a *textural class* according to the triangular diagram of Figure 2.3, which summarizes all the possible combinations in which the sand, silt and clay may occur. The Soil Survey of England and Wales has established 11 major textural classes based upon the British Standards system of particle-size grading (Figure 2.2). The field assessment of texture can be checked by carrying out a complete analysis of the particle-size distribution by physical means.

Particle-size analysis in the laboratory
The success of the method relies upon the complete disruption of soil aggregates and the addition of chemicals which ensure the dispersion of the soil colloids in water. Full details of the methods employed are given in standard texts—for example, Black (1965) and Loveday (1974). The coarse sand particles are separated by sieving; fine sand, silt and clay are separated by making use of the differences in their settling velocities in the suspension. The principle of the latter technique is as follows:

A rigid particle falling freely through a liquid of a lower density will attain a constant velocity when the force opposing movement is equal and opposite to the

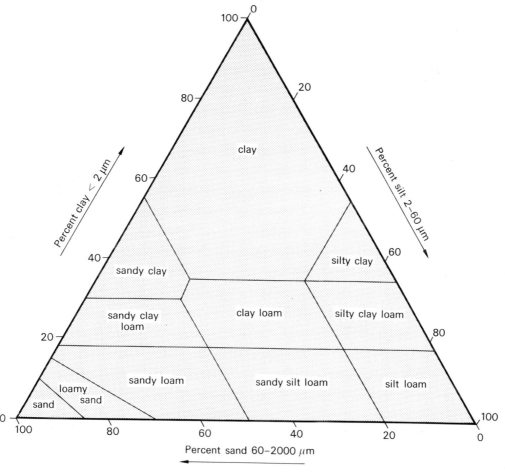

Figure 2.3. Triangular diagram of soil textural classes (after SSEW, 1974).

force of gravity acting on the particle. The frictional force acting vertically upward on a spherical particle is calculated from Stoke's law. The net gravitational force acting downwards is equal to the weight of the sub-merged particle. At equilibrium, these expressions can be combined to give an equation for the terminal settling velocity v as

$$v = \frac{2}{9}\frac{g}{\eta}(\rho_p - \rho)r^2 \qquad (2.1)$$

where g = acceleration due to gravity
$\quad \eta$ = coefficient of viscosity of the liquid (water)
$\quad \rho_p$ = density of the particle

ρ = density of water
r = radius of the particle

When all the constants in this equation are collected into one term K, we derive the simple relationship

$$v = \frac{h}{t} = Kr^2 \qquad (2.2)$$

where h is the depth, measured from the liquid surface, below which all particles of radius r will have fallen in time t. To illustrate the use of equation 2.2, we may note that all particles >2 μm diameter settling in a suspension at 20°C will have fallen below a depth of 10 cm in 8 hours. Thus, by sampling the weight per unit volume of the suspension at this depth after 8 hours, the amount of clay present can be calculated. The clay content is normally measured with a Bouyoucos hydrometer placed in the suspension, or calculated from the loss of weight of a bulb of known volume immersed in the suspension—the *plummet balance* method, illustrated in Figure 2.4.

(a) the clay and silt particles are not smooth spheres—they have irregular plate-like shapes; and
(b) the density of the particle varies with its mineralogy (see section 2.3). In practice, we take an average value for ρ_p of 2.65, and speak of the 'equivalent spherical diameter' of the particle size being measured.

The result of the analysis is expressed as the weights of the individual fractions per 100 g of oven-dry soil (fine earth only). Oven-dry soil is soil dried to a constant weight at 105°C. By combining the coarse and fine sand fractions, the soil may be represented by one point on the triangular diagram of Figure 2.3. Alternatively, by the stepwise addition of particle-size percentages, graphs of cumulative percentage against particle diameter of the kind shown in Figure 2.1 can be obtained.

2.2 THE IMPORTANCE OF SOIL TEXTURE

Soil scientists are primarily interested in the texture of the fine earth of the soil. Nevertheless, in some soils the size and abundance of the stones cannot be ignored, since they can have a marked influence on the soil's suitability for agriculture. As the stone content increases, the soil will hold less water than a stoneless soil of the same fine-earth texture, so that crops become more susceptible to drought. Conversely, the free drainage of such soils may be advantageous under wet conditions. Stoniness also determines the ease, and to some extent the cost of cultivation, as well as the abrasive effect of the soil on tillage implements.

Texture is one of the most stable of properties, and is a useful index of several other properties which determine the soil's agricultural potential. The fine and medium-textured soils, such as the clays, clay loams, silty clay loams and sandy silt loams, are generally the more desirable because of their superior retention of nutrients and water. Conversely, where high rates of infiltration and good drainage are required, as in irrigation schemes, sandy or coarse-textured soils are preferred. In farming parlance, clay soils are described as 'heavy' and sandy soils as 'light': this does not refer to their weight per unit volume, which may be greater

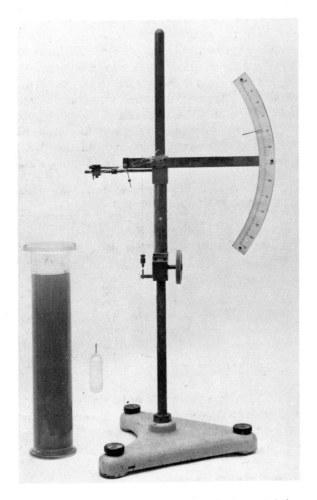

Figure 2.4. A settling soil suspension and plummet balance (courtesy of J. Loveday).

Great accuracy in measuring the weight per unit volume of the suspension is not warranted because of the simplifying assumptions made in the calculation of settling velocity. In particular, it should be noted that:

for a sandy soil than a clay, but to the power required to draw implements through the soil.

Texture has a pronounced effect on soil temperature. Clays hold more water than sandy soils, and the presence of water modifies considerably the heat requirements of the soil—firstly, because its specific heat capacity is 3 to 4 times that of the soil solids, and secondly, because of the quantity of latent heat either absorbed or evolved during a change in the physical state of water, for example, from ice to liquid. Thus it is found that the temperature of wet clay soils responds more slowly than that of sandy soils to changes in air temperature in spring and autumn.

2.3 MINERALOGY OF THE SAND AND SILT FRACTIONS

Simple crystalline structures

Sand and silt consist almost entirely of the resistant residues of *primary* rock minerals, although small amounts of secondary minerals—salts, oxides and hydroxides also occur. The primary rock minerals are predominantly *silicates* which have a crystalline structure based upon a simple unit, the *silicon tetra-*

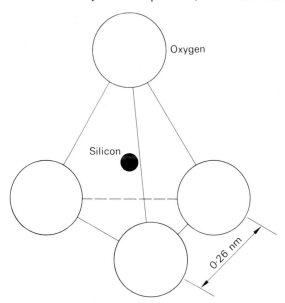

Figure 2.5. Diagram of a silicon tetrahedron.

hedron SiO_4^{4-} (Figure 2.5). An electrically-neutral crystal is formed when cations such as Al^{3+}, Fe^{3+}, Fe^{2+}, Ca^{2+}, Mg^{2+}, K^+ and Na^+ become covalently bonded to the oxygen atoms. The surplus valencies of the oxygens in the SiO_4^{4-} group are then satisfied.

Coordination number

The packing of the oxygen atoms, which are by far the largest of the more abundant elements in the silicates, determines the crystalline dimensions. In quartz, for example, oxygen occupies 98.7 per cent and silicon only 1.3 per cent of the mineral volume. The size ratio of Si to O is such that four oxygens can be packed around one silicon. Larger cations can accommodate more oxygens, the ratio of the ionic radius of the cation to the ionic radius of oxygen being related to the *coordination number* of the cation, examples of which are given in Table 2.1. Aluminium, which has a cation/oxygen

Table 2.1. Cation/oxygen radius ratios and coordination numbers for common elements in the silicate minerals.

Radius of cation/radius of oxygen			
Coordination No. 3	Coordination No. 4	Coordination No. 6	Coordination No. 8 and greater
B 0.15	Be 0.26	Al 0.43	Na 0.74
		Ti 0.48	Ca 0.80
	Si 0.30	Fe^{3+} 0.51	
	Mn^{4+} 0.39	Mg 0.59	Sr 0.96
		Fe^{2+} 0.63	
		Zr^{4+} 0.69	K 1.01
		Mn^{2+} 0.69	Ba 1.08

(after Marshall, 1964)

radius ratio close to the maximum for coordination number 4 and the minimum for coordination number 6 can exist in either fourfold (IV) or sixfold (VI) co-ordination.

Isomorphous substitution

Elements of the same valency and coordination number frequently substitute for one another in a silicate structure—a process called *isomorphous substitution*. The structure remains electrically neutral. However, when elements of the same coordination number but different valency are exchanged, there is an imbalance of electric charge. The most common substitutions are

Mg^{2+} (VI) for Fe^{3+} (VI) or Al^{3+} (VI), and Al^{3+} (IV) for Si^{4+} (IV). The excess negative charge is neutralized by the attraction of additional cations such as K^+, Na^+, Mg^{2+} or Ca^{2+} to the crystal lattice, or by structural arrangements which allow an internal compensation of charge (see chlorites and smectites).

More complex crystalline structures

CHAIN STRUCTURES

These are represented by the *pyroxene* and *amphibole* groups of minerals, which collectively make up the ferromagnesian minerals. In the pyroxenes, each silicon tetrahedron is linked to adjacent tetrahedra by the sharing of two out of three of the basal oxygen atoms to form a single extended chain (Figure 2.6). In the

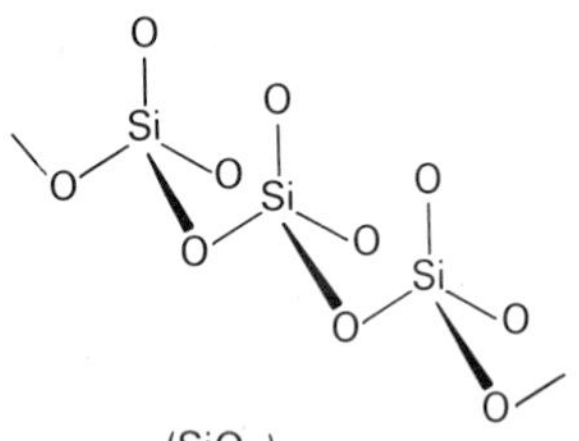

$(SiO_3)_n$

Fig. 2.6. Single chain structure of silicon tetrahedra (after Moore and Moore, 1976).

amphiboles, a double parallel chain structure is formed when all three of the basal oxygen atoms of a silicon tetrahedron are shared with adjacent tetrahedra (Figure 2.7).

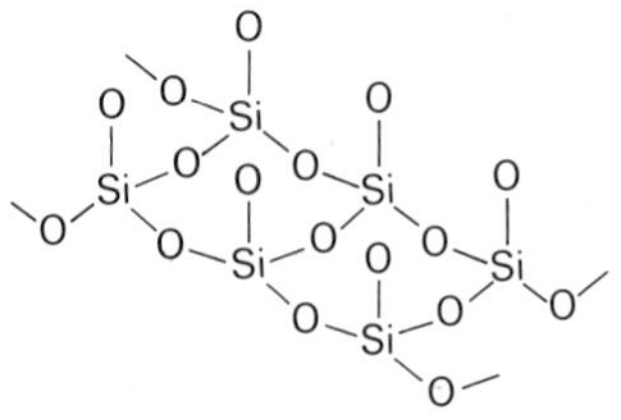

$(Si_2O_5)_n$

Figure 2.7. Double chain structure of silicon tetrahedra (after Moore and Moore, 1976).

SHEET STRUCTURES

It is easy to visualize the two-dimensional extension of a single chain structure to form a sheet structure, which

is characteristic of the *micas, chlorites* and the *clay minerals* (section 2.4). The sheet displays a pattern of hexagonal holes when seen in plan view (Figure 2.8). The mica-type structure is built up by the combination of *two* silica sheets with *one* sheet of aluminium atoms

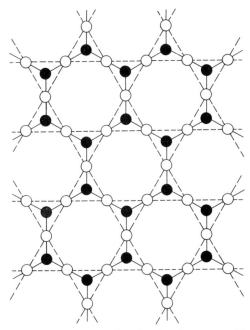

Fig. 2.8. A silica sheet in plan view showing the pattern of hexagonal holes (after Fitzpatrick, 1971).

in sixfold coordination with O and OH. (Because the unhydrated proton is so small, OH occupies virtually the same space as O.) The unit structure of the alumina sheet is the *aluminium octahedron* (Figure 2.9). Aluminium atoms normally occupy only two-thirds of the available octahedral positions in this sheet, giving rise to the characteristic mineral structure of *gibbsite*, $[Al_2OH_6]_n$. If Mg is present instead of Al, however, all the available octahedral positions are filled, giving rise to the structure of *brucite* $[Mg_3(OH)_6]_n$. The complete, plate-like 2:1 lattice constitutes a crystalline layer or *lamella*, and the regular stacking in the vertical (or *C* direction) of such lamellae makes up a *crystal*. This is illustrated in Figure 2.10.

Additional structural complexity is introduced by isomorphous substitution. In the mica minerals *biotite* and *muscovite*, one-quarter of the tetrahedral Si is

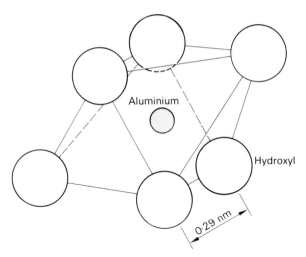

Figure 2.9. Diagram of an aluminium octahedron.

replaced by Al resulting in a net negative charge of 2 equivalents* per unit cell, which has the composition:

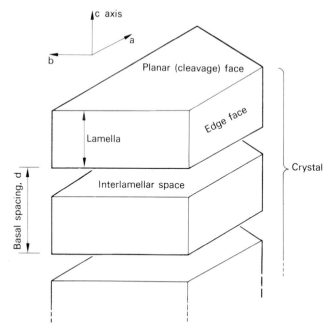

Figure 2.10. General structure of a clay crystal.

* One equivalent of charge is the charge of an electron $\times$ Avogadro's Number (6.0×10^{23}); it is also equal to the charge carried by 1 g of H^+ ions.

$[(OH)_4 \, (Al_2Si_6)^{IV} \, (Mg \, Fe)_6^{VI} \, O_{20}]^{2-} \, 2K^+$ for biotite, and

$[(OH)_4 \, (Al_2Si_6)^{IV} \, Al_4^{VI} \, O_{20}]^{2-} \, 2K^+$ for muscovite.

Note that the negative charge is neutralized by K^+ ions held in the spaces formed by the juxtaposition of the hexagonal holes of adjacent silica sheets (Figure 2.11). The positioning of the K^+ ions makes the bonding between the lamellae very strong.

The most complex of the two dimensional structures belongs to the *chlorites* which have a brucite layer sandwiched between two mica layers. In the type-mineral *chlorite*, the negative charge of the two biotite layers is neutralized by a positive charge in the brucite, developed due to the replacement of two-thirds of the Mg atoms by Al (Figure 2.12). Predictably, the bonding between lamellae is strong, but the high content of Mg renders this mineral susceptible to dissolution in acidic solutions. For the same reason, and also because of its ferrous iron content, biotite is much less stable than muscovite.

Three dimensional structures

The most important silicates in this group are *silica* and the *feldspars*. Silica minerals consist entirely of polymerized silicon tetrahedra of general composition $(SiO_2)_n$. Silica occurs as the residual mineral *quartz*, which is very inert, and as a secondary mineral precipitated after the hydrolysis of more complex silicates, which when metamorphosed appears as flint or chert. It also occurs as amorphous or microcrystalline silica of biological origin, the fossilized remains of the diatom, a minute aquatic organism, being almost pure silica. Silica absorbed by terrestial plants (grasses and hardwood trees in particular) is also returned to the soil as microscopic, opaline structures called *phytoliths*, or as amorphous material in the ash.

Quartz or flint fragments of greater than colloidal size are very insoluble, and hence are abundant in the sand and silt fractions of many soils. The *feldspars*, on the other hand, are chemically more reactive and are rarely of significance in the sand fraction of mature soils. Their structure consists of a three-dimensional framework or polymerized silicon tetrahedra in which some of the Si is replaced by Al. The cations balancing the excess negative charge are all of high coordination

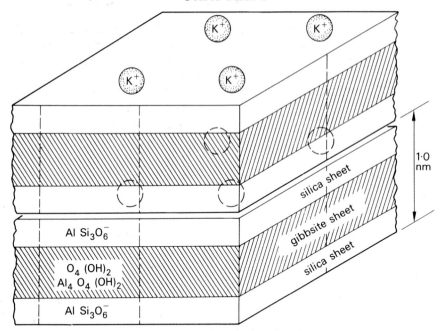

Figure 2.11. Structure of muscovite mica—a 2 : 1 non-expanding, layer-lattice mineral.

Figure 2.12. Structure of chlorite—a 2 : 2 or mixed layer mineral.

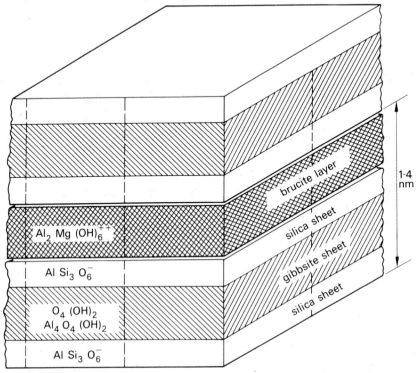

number, such as K^+, Na^+ and Ca^{2+}, and less commonly Ba^{2+} and Sr^{2+}. The range of composition encountered is shown in Figure 2.13. Details of the structure,

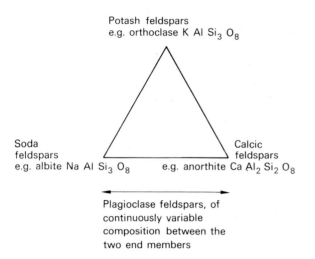

Potash feldspars
e.g. orthoclase $K\,Al\,Si_3\,O_8$

Soda
feldspars
e.g. albite $Na\,Al\,Si_3\,O_8$

Calcic
feldspars
e.g. anorthite $Ca\,Al_2\,Si_2\,O_8$

Plagioclase feldspars, of continuously variable composition between the two end members

Figure 2.13. Type minerals of the feldspar group.

composition and chemical stability of the feldspars are given in specialist texts by Marshall (1964) and Loughnan (1969).

2.4 MINERALOGY OF THE CLAY FRACTION

For many years the small size of colloidal particles prevented scientists from elucidating their mineral structure. It was thought that the clay fraction consisted of inert mineral fragments enveloped in an amorphous gel of hydrated *sesquioxides* ($Fe_2O_3 . n\,H_2O$ and $Al_2O_3 . n\,H_2O$) and *silicic acid* ($Si(OH)_4$). The surface gel was *amphoteric*, the balance between acidity and basicity being dependent on the pH of the soil. Between pH 5 and 8, the surface was usually negatively charged (proton-deficient), which could account for the observed cation exchange properties of the soil (section 2.5).

During the 1930s, however, the crystalline nature of the clay minerals was established unequivocally by the technique of X-ray diffraction. Most of the minerals were found to have a plate-like structure similar to that of the micas and chlorite. The various mineral groups were identified from their characteristic *basal spacings* (Figure 2.10), as measured by X-ray diffraction, and their unit cell compositions deduced from elemental analyses. Subsequent studies using the scanning and transmission electron microscopes have fully confirmed the conclusions of the early work. However, it is known that *accessory minerals*—the weathered residues of resistant primary minerals that have been comminuted to colloidal size—also occur in the clay fraction, as well as variable amounts (~ 20 percent by weight) of *amorphous material*, which is of considerable importance since it can occur as a thin covering on the surface of the clay minerals.

The crystalline clay minerals

These minerals are hydrous aluminosilicates with a layer-lattice structure, except for the small group with chain structures similar to amphiboles (the sepiolite-attapulgite series). They can be inherited from the parent material as partially-altered fragments of the primary layer silicates—the micas and chlorites, or may be formed during pedogenesis—the kaolinites, hydrous micas and smectites.

KAOLINITES

The basic structure of this group is a 1:1 layer-lattice formed by the sharing of some oxygen atoms between a silica sheet and a gibbsite sheet. The unit cell composition is $Si_2^{IV} O_5 Al_2^{VI} OH_4$. The two most common members of the group found in soils are *kaolinite* and *halloysite*.

Kaolinite has a basal spacing fixed at 0.7 nm due to hydrogen-bonding between the hydrogen and oxygen atoms of adjacent lamellae (Figure 2.14). The lamellae are stacked fairly regularly in the C direction to form crystals from 0.05 to 2 μm thick, the larger crystals occurring in relatively pure deposits of China Clay. The crystals are hexagonal in plan view and usually larger than 0.2 μm in diameter, as shown in the electron-micrograph of Figure 2.15. Halloysite has the same structure as kaolinite, with the addition of 2 or 4 molecules of water, per unit cell, which is located in the interlamellar space. The presence of this hydrogen-bonded water so alters the distribution of stresses

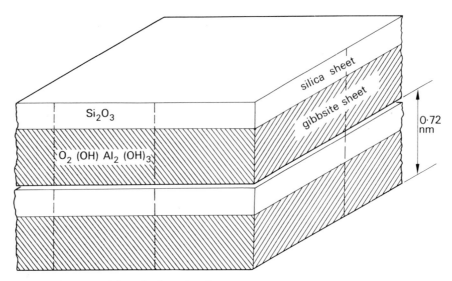

Figure 2.14. Structure of kaolinite—a 1 : 1 layer-lattice mineral with hydrogen-bonding between lamellae.

within the mineral lattice that the layers curve to form a tubular structure.

There is some isomorphous substitution of Al for Si in kaolinite which produces a very low net negative charge of *c*. 0.005 equivalents per unit cell. In addition, a negative charge can develop due to unsatisfied oxygen and hydroxyl valencies at the edges of the crystals (Figure 2.16) which at a pH >9 can contribute as much 0.008 milli-equivalents per m² of edge area (*cf.* Table 2.3).

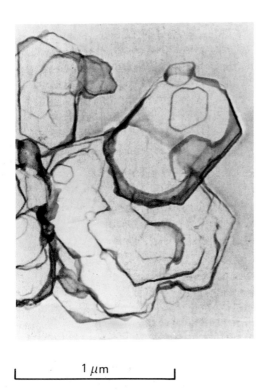

Figure 2.15. Electron micrograph of kaolinite crystals.

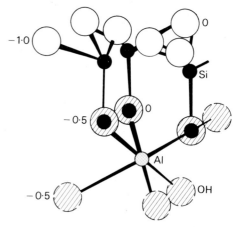

Figure 2.16. Charges on the edge face of kaolinite at high pH (after Hendricks, 1945).

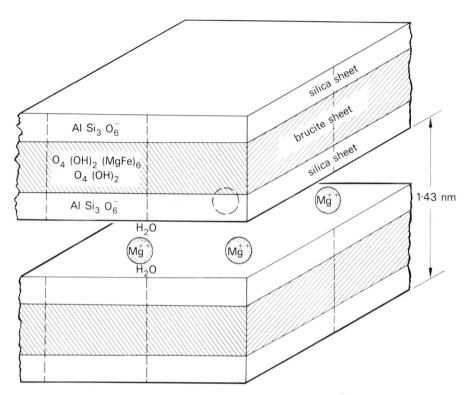

Figure 2.17. Structure of vermiculite—basal spacing fixed in the presence of partially hydrated Mg^{2+} ions.

HYDROUS MICAS

These are closely related to the micas (section 2.3). The *vermiculites* are analogues of biotite and the *illites* of muscovite. The hydrous micas have less substitution of Al for Si than the micas, resulting in an average negative charge of 1.3 equivalents per unit cell, and have higher contents of crystalline water.

In the *vermiculites*, interlamellar K^+ is replaced by Ca^{2+} and Mg^{2+}. These cations are more strongly hydrated than K^+, so that in the interlamellar space they are associated with a bimolecular layer of oriented water molecules which stretches the basal spacing from the characteristic 1.0 nm of a mica to 1.4–1.5 nm (Figure 2.17). On heating and dehydration, the lattice collapses to a 1.0 nm spacing, so that the vermiculites show limited reversible swelling.

In *illites*, K^+ remains as the major cation although it is thought that, to account for the extra water content compared to the original mica, some of the inter-lamellar K^+ is replaced by H_3O^+ ions, or there is some interstratification of vermiculite and illite layers within the crystal. The interlamellar bonding is much weaker than in muscovite mica, and the stacking of the layers is therefore less orderly. Nevertheless, pure illites have a non-expanding lattice with a basal spacing of 1.0 nm.

SMECTITES

Minerals of the smectite group have a 2:1 structure similar to that of the hydrous micas, but exhibit a much wider range of composition, with isomorphous substitutions occurring in both the tetrahedral and octahedral sheets. All members of the group have a net negative charge which is balanced by hydrated cations (Ca^{2+}, Mg^{2+}, K^+ and Na^+) situated in the interlamellar spaces with no high degree of orientation. Accordingly, the interlamellar bonding is weak, and depending on the hydration of the cations present, the

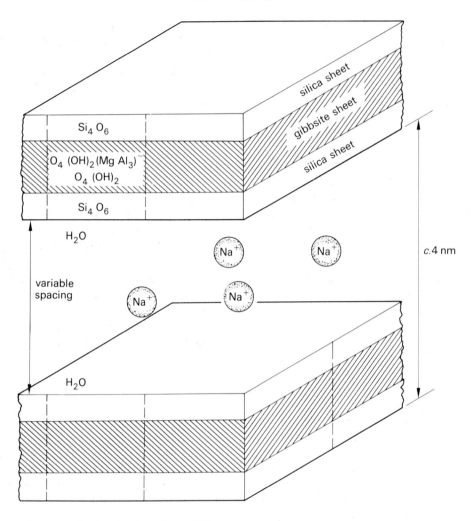

Figure 2.18. Structure of montmorillonite—basal spacing variable depending on the dominant exchangeable cation.

basal spacing may vary from 1 to >4 nm (Figure 2.18). The Ca^{2+} forms, predominant in soils, have a basal spacing of 1.9 nm when fully hydrated. The negative charge averages 0.66 equivalents per unit cell.

The most important smectites in soils are the minerals *montmorillonite*, *beidellite* and *nontronite*. Because the interlamellar regions are readily accessible to hydrated cations and polar molecules, these clays are sometimes called expanding-lattice clays. The stacking of the lamellae is very irregular and the average crystal size is therefore much smaller than in the hydrous micas and kaolinites. The planar interlamellar surfaces provide a large internal area which augments the external cleavage face area so that the total surface area of a smectite clay is very large indeed (see Table 2.2).

MIXED LAYER MINERALS
Regular and irregular interstratification of clay minerals produces mixed layer minerals. The chlorites, for example, are formed by the *regular* interstratification of brucite or gibbsite layers and biotite mica. The basal spacing is a consistent 1.4 nm. With the improvement

of X-ray diffraction techniques, *irregular* interstratification has been found to occur much more frequently in clay minerals separated from soil than was previously believed. Mixed layers of the 2:1 minerals are most common, especially intergrades of vermiculite and illite, but examples of 1:1 mixed layer minerals also occur, such as *anauxite*, which is formed by the random insertion of a double silica sheet between kaolinite lamellae.

Accessory minerals and amorphous materials
The only salt of importance found in the clay fraction is *calcite* ($CaCO_3$) which can accumulate in soils formed on chalk and limestone, and in high pH soils of arid regions. The accessory minerals are predominantly *free oxides*—compounds in which a single cation species is coordinated with oxygen, which include crystalline and amorphous forms of silica, iron oxide, aluminium oxide, manganese dioxide (MnO_2) and anatase (TiO_2).

SILICA
The various forms of silica have been described (section 2.3). Silicic acid in solution has a marked tendency to polymerize as the pH rises so that colloidal silica gels can occur as cementing agents in the lower B and C horizons of leached soils. Amorphous silica also occurs in soils formed on recent deposits of volcanic ash, as has been observed in Japan, New Zealand and Hawaii. Indeed, the bulk of the clay fraction in such soils may consist of a microcrystalline compound of silica and alumina called *allophane*. Allophanes have extremely high negative charge densities because approximately one-third of the Si atoms is replaced by tetrahedral Al.

IRON OXIDES
The yellow to reddish-brown colour of the iron oxides is very obvious in many highly weathered soils of the Tropics, and in the rusty mottling of soils which undergo periodic waterlogging. The rust mottles are frequently associated with small black accumulations (1–2 mm diameter) of MnO_2.

Iron oxide occurs most commonly as the hydrated crystalline minerals *goethite* ($\alpha - FeOOH$), which is the normal weathering product, and *lepidocrocite* ($\gamma - FeOOH$) which is formed by the oxidation of ferrous (Fe^{2+}) iron in the presence of organic matter. Lepidocrocite slowly reverts to goethite on ageing, and goethite is changed to *hematite* ($\alpha - Fe_2O_3$) at the high temperatures prevailing in the surface of tropical soils. Hydrated *magnetite* ($Fe_3O_4 . nH_2O$) contributes to the blue-grey colour of waterlogged soils.

ALUMINIUM OXIDES
The aluminium oxides have a non-distinctive greyish-white colour which is easily masked in soils except when large concentrations occur as in bauxite ores. The principal crystalline form is the hydrated oxide *gibbsite* $\gamma - Al(OH)_3$; deposits of amorphous $Al(OH)_3$ also occur in the interiamellar spaces of clay minerals, especially the vermiculites, and as surface coatings on clay particles generally. This amorphous material slowly crystallizes to form gibbsite. Both aluminium and iron oxides accumulate under intense weathering conditions in the Tropics (section 9.2), and their presence has a profound effect on the physical and chemical properties of the soils formed.

2.5 SURFACE AREA AND SURFACE CHARGE

Specific surface
We can demonstrate that the smaller the size of an object, imagined as an ideal sphere, the greater the ratio of its surface to volume. Thus, when the radius $r = 1$ mm,

$$\frac{\text{Surface area}}{\text{volume}} = \frac{4\pi r^2}{4/3\pi r^3} = 3 \text{ mm}^{-1}$$

and when $r = 0.001$ mm (or 1 micron),

$$\frac{\text{Surface area}}{\text{volume}} = \frac{4\pi \times 10^{-6}}{4/3\pi \times 10^{-9}} = 3 \times 10^3 \text{ mm}^{-1}$$

For convenience, the *specific surface* of a soil particle is defined as the surface area per unit weight, which implies a constant particle density. Representative values for the specific surfaces of sand, silt and clay-size minerals are given in Table 2.2.

Table 2.2. Specific surface areas according to mineral type and particle size.

Mineral or size class	Specific surface ($m^2 g^{-1}$)	Method of measurement
Coarse sand	0.01	Low temperature N_2
Fine sand	0.1	adsorption (B.E.T.)
Silt	1.0	
Kaolinites	5–100	
Hydrous micas (illites)	100–200	B.E.T.
Vermiculites and mixed layer minerals	300–500	Ethylene glycol
Montmorillonite (Na-saturated)	700–800	adsorption
Allophanes	200–500	
Amorphous and hydrated iron and aluminium oxides	100–300	B.E.T.

(after Fripiatt, 1965)

ADSORPTION

When a boulder is reduced to small rock fragments by weathering, part of the energy put in is conserved as free energy associated with the new particle surfaces. The higher the specific surface of a substance, therefore, the greater is its surface free energy. Since free energy tends to a minimum (Second Law of Thermodynamics), the surface energy is reduced by the work done in attracting substances to the surface, which gives rise to the phenomenon of surface *adsorption*. The adsorption of nitrogen gas at very low temperatures on thoroughly dry clay is used to measure the specific surface of the the clay (the Branauer, Emmet and Teller or B.E.T. method) since the adsorbed N_2 forms a monomolecular layer with an area of $0.16 \ nm^2$ per molecule. Where the mineral has internal surfaces, as in montmorillonite, an additional measurement is required because the non-polar N_2 molecules do not penetrate into the interlamellar spaces; the polar molecule of ethylene glycol does, and it gives a measure of the *internal* and *external* surfaces. In the absence of an internal surface, as in kaolinite, the B.E.T. and ethylene glycol methods should give identical results. As shown in Table 2.2, the total specific surface of Na-montmorillonite approaches $800 \ m^2 g^{-1}$, which is about the maximum

for a completely dispersed clay of 1 nm lamellar thickness.

SURFACE CHARGES

In addition to the adsorption of neutral molecules, cations and anions are adsorbed at clay and oxide surfaces as a result of:
(a) the permanent negative charges of the clays due to isomorphous substitution; and
(b) the amphoteric properties of the edge faces of clays and the surfaces of hydrated iron and aluminium oxides.

This behaviour is discussed more fully in Chapter 7. Suffice to say here that the norm is one of cation attraction and anion repulsion. Cations adsorbed on the external and internal clay surfaces that are freely *exchangeable* to cations in solution give rise to the *cation exchange capacity* (*CEC*), which is expressed in milli-equivalents (me) per 100 g of soil or clay (Table 2.3). The *CEC* is quite variable within and between

Table 2.3. Cation exchange capacities and surface charge densities of the clay mineral groups.

Clay mineral group	CEC me $100 \ g^{-1}$	Surface charge density (Γ) me m^{-2}
Kaolinites	3–20	2–6×10^{-3}
Illites	10–40	1–2×10^{-3}
Smectites	80–120	1–1.5×10^{-3}
Vermiculites	100–150	3×10^{-3}

mineral groups, and not always the value expected from the number of equivalents of unbalanced negative charge per unit cell. Of the hydrous micas, for example, the *CEC* of the illites is low because much of the charge is neutralized by interlamellar K^+ which is *non-exchangeable*; on the other hand, the interlamellar Ca^{2+} and Mg^{2+} of the soil vermiculites are exchangeable and the *CEC* is high.

The ratio *CEC*/specific surface defines the *surface charge density* (Γ) of the clay in me m^{-2}, the appropriate values of which are given for the different minerals in Table 2.3. Despite the large range in charge per unit weight (the *CEC*), the values of Γ are reasonably similar, suggesting that the clay crystals cannot acquire an indefinite number of negative charges and remain stable. Kaolinite, which has very little isomorphous substitution, exists as large crystals and so has a much lower specific surface than montmorillonite; on the other hand, the large kaolinite crystals have a relatively greater edge area at which additional negative charges can develop and so augment the total surface charge density of the clay.

2.6 SUMMARY

There is a continuous distribution of particle sizes in soil from boulders and stones down to clay minerals less than 2 μm in equivalent diameter. The material which passes through a 2 mm sieve—the *fine earth*, is divided into *sand*, *silt* and *clay*, the relative proportions of which determine the soil *texture*. Texture is a property that changes only slowly with time, and is an important determinant of the soil's response to water—its stickiness, mouldability and permeability; its capacity to retain cations, and its rate of adjustment to ambient temperature changes.

Cation retention, and the adsorption of water and solutes, also depend on the kind of clay minerals in the soil. These occur as flat crystals, often hexagonal in shape, in which the individual layers or lamellae are made up of one *silica sheet* $[SiO_2]_n$ combined with one *gibbsite sheet* $[Al_2(OH)_6]_n$—the *kaolinite group*; or one gibbsite sheet sandwiched between two silica sheets—the *mica group*, which includes *illite*, *vermiculite* and *montmorillonite*. Because of lattice substitutions, these crystalline clays have a permanent negative charge which is neutralized by cations adsorbed from solution, giving rise to a *cation exchange capacity (CEC)*. The *CEC* varies from 3 me per 100 g clay in the kaolinites to 150 me per 100 g in some vermiculites. The edge faces of clay minerals have amphoteric properties, as

do crystalline and amorphous forms of the oxides $Al_2O_3 . nH_2O$ and $Fe_2O_3 . nH_2O$ which can occur as discrete particles or as surface coatings in the clay fraction of the soil. They are positively charged at low pH and negatively charged at high pH.

The smaller the particle, the larger the surface area to volume ratio—the *specific surface*, and the greater the potential for adsorption. When the interlamellar surfaces of the clay crystal are accessible to water and cations, the total surface area is the sum of the internal and external areas, and approaches 800 m^2 g^{-1} for a fully dispersed Na-montmorillonite. However, when the crystal lamellae are strongly bonded to one another, as in kaolinite, the surface area comprises only the external surface and may vary from 5 to 100 m^2 g^{-1} depending on the crystal size.

REFERENCES

BLACK C.A. (1965) (Ed.) Methods of soil analysis. Part I. *American Society of Agronomy Monograph No. 9.*

LOUGHNAN F.C. (1969) *Chemical Weathering of the Silicate Minerals.* American Elsevier, New York.

LOVEDAY J. (1974) (Ed.) Methods for analysis of irrigated soils. *Commonwealth Bureau of Soils, Technical Communication No. 54.*

MARSHALL C.E. (1964) *The Physical Chemistry and Mineralogy of Soils.* I. *Soil Materials.* Wiley, New York.

FURTHER READING

FRIPIAT J.J. (1965) Surface chemistry and soil science, in *Experimental Pedology* (Eds. E.G. Hallsworth and D.V. Crawford). Butterworth, London.

GRIM R.E. (1968). *Clay Mineralogy*, 2nd Ed. McGraw-Hill, New York.

HENDRICKS S.B. (1945) Base exchange of crystalline silicates. *Industrial and Engineering Chemistry* 37, 625–630.

MITCHELL B.D., FARMER V.C. & McHARDY W.J. (1964) Amorphous inorganic material in soils. *Advances in Agronomy* 16, 327–383.

Chapter 3
Soil Organic Matter

3.1 ORIGIN OF SOIL ORGANIC MATTER

The carbon cycle

The organic matter of the soil arises from the debris of green plants, animal residues and excreta which are deposited on the surface and admixed to a variable extent with the mineral component. The dead organic matter is colonized by a variety of *soil organisms* which derive energy for growth from the oxidative decomposition of complex organic molecules. During decomposition, essential elements are converted from organic combination to inorganic forms such as ammonium (NH_4^+), phosphate ($H_2PO_4^-$) and sulphate ($SO_4^=$)—a process called *mineralization*. Mineralization, and the release of some of the organic carbon as CO_2, is vital for the growth of succeeding generations of green plants. A proportion of this nutrient release is rendered temporarily unavailable for plant growth due to its incorporation into the cell substance of the soil organisms—a process called *immobilization*. The various interlocking processes of synthesis and decomposition by which carbon is circulated through the biosphere comprise the *carbon cycle* (Figure 3.1).

Inputs of plant and animal residues

The annual return of plant and animal residues to the soil varies greatly with the climatic region and the type of vegetation or land use. Some average figures for annual litter fall, in terms of its organic carbon content, are given in Table 3.1. These values do not include the contributions from root death and decay, estimated at one-third to one-half of the leaf fall, and timber fall in the forests, which can be at least equal to the rate of leaf fall, although not providing a substrate as readily

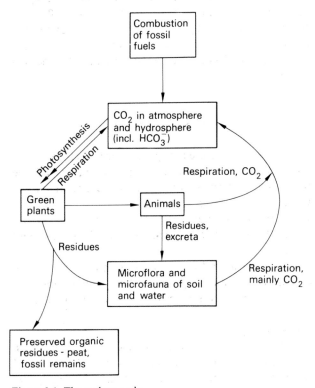

Figure 3.1. The carbon cycle.

Table 3.1. Annual rate of litter return to the soil.

Land use or vegetation type	Organic C
	(t ha^{-1})
Alpine and arctic forest	0.1
Arable farming (cereals)	1–2
Temperate grassland	2–4
Coniferous forest	1.5–3
Deciduous forest	1.5–4
Tropical rainforest (Colombia)	4–5
Tropical rainforest (West Africa)	10

To convert from organic C to organic matter, divide these figures by 0.6.

decomposable. A comparable figure for the amount of dung produced by dairy cows grazing highly productive pastures is 3.5 to 4.5 tonnes ha^{-1} an^{-1}.

Composition of plant litter

Plant cells are made up primarily of carbohydrates, proteins and fats, plus smaller amounts of organic acids, lignins, waxes and resins. The bulk of the material is carbohydrate of which sugars and starch are rapidly decomposed, and hemi-cellulose and cellulose less rapidly decomposed, to provide metabolic energy for the soil organisms.

Proteins are readily decomposed to amino acids, and depending on the ratio of C:N in the plant tissue, all or part of the amino-N will be incorporated into the protein of soil organisms. When the C:N ratio is greater than 25, there is generally no amino-N that is surplus to the growth demands of the organisms. At C:N ratios <25, however, some of the amino-N is released in inorganic form and *net mineralization* of the plant nitrogen occurs. As can be seen from Table 3.2, the C:N ratio of fresh litter is highly variable

Table 3.2. C : N ratios of freshly-fallen litter and of soils.

	Range	Mean
Litter		
Herbaceous legumes	—	20
Cereal straw	40–120	80
Tropical forest species (Colombia)	27–32	30
Temperate hardwoods (elm, ash, lime, oak, birch)	25–44	35
Scots pine	—	91
Soil		
Soils from 63 sites in the U.S.A.	7–26	12.8
Rothamsted: old woodland	—	9.5
Old pastures, fertilized and unfertilized	11–12	11.8
Old arable, manured and unmanured	8–10	9.4

among plant species, but as the organic matter passes through successive cycles of decomposition in the soil, the C:N ratio gradually narrows, and we find the C:N ratio of well drained soils of pH ~7 to be very close to 10. Exceptions occur with poorly drained soils or those on which mor humus forms (section 3.3).

The plant organic acids are readily decomposed: not so the fats, waxes and resins which can persist in the soil for some time. Similarly the lignins, which are complex phenolic polymers, are more stable than the carbohydrates and accumulate relative to these constituents during the decomposition of fresh residues, as

Table 3.3. Changes in the major constituents of rye straw during aerobic decomposition.

Constituent	At the start	After 2 months
	(per cent)	
Total dry matter (ash-free)	100	58
Cellulose	41.5	18.3
Lignin	22.5	20.0
Protein	1.2	3.4

(after Waksman, 1938)

indicated in Table 3.3. Lignins attain significant proportions in the dry matter of cereal straw (10–20%) and wood (20–30%). Their degradative products, the monocyclic phenols, probably serve as precursors for the synthesis of complex new compounds by the soil organisms.

3.2 THE SOIL ORGANISMS

The soil organisms are grouped, according to size, into:
The macrofauna—vertebrate animals mainly of the burrowing type such as moles and rabbits, which live wholly or partly underground;
The mesofauna—small invertebrate animals representative of the phyla Arthropoda, Annelida, Nematoda and Mollusca;
The microfauna and *microflora* (collectively the *microorganisms*), comprising five major subgroups—the bacteria, fungi, actinomycetes, algae and protozoa.

BIOMASS

The weight of the organisms is referred to as *biomass*. For the macro- and mesofauna, which can be separated from inert soil material, the biomass is usually expressed as liveweight, in kg ha^{-1} to a certain depth (section 4.5). As we shall see, however, the micro-organisms are intimately mixed with the dead organic matter in the soil and, being very small, are difficult to isolate for counting and weighing. Indirect methods can be used to estimate the microbial biomass, based on the flush of decomposition and evolution of CO_2 following the reinoculation of a soil sterilized by fumigation (section 3.5). The microbial biomass is therefore expressed as a weight of organic C, in kg ha^{-1}, or as a percentage of the total soil organic matter. In arable soils, the microbial biomass is about 2 per cent of the soil organic matter, rising to about 3 per cent in grassland and woodland soils. These percentages correspond to weights of organic C ranging from 0.5 to 2.0 t ha^{-1} to a depth of 25 cm.

The micro-organisms

BACTERIA

These minute organisms can be as small as 1 μm in length and 0.2 μm in breadth and therefore live in water films around soil particles in all but the finest pores. They can be motile or non-motile, coccoid (round) or rod-shaped (Figure 3.2), and can reproduce

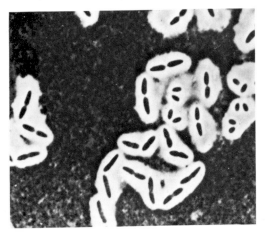

Figure 3.2. Soil bacteria—*Azotobacter* sp. (after Hepper, 1975).

very rapidly in the soil under favourable conditions—as little as 8–24 hr for the production of two daughter organisms from a single parent by fission. Accordingly,

their number in the soil is enormous provided that living conditions, especially the food supply, are suitable. Bacterial numbers are estimated by observing the growth of colonies on special nutrient media which have been inoculated with drops of a very dilute soil suspension (the *dilution plate method*), or by direct microscopic observation of suitably stained thin soil sections, recording all the micro-organisms—living, dead and those which have formed spores. Estimates of between 1 and 4×10^9 organisms g^{-1} of soil are obtained by the latter method.

The bacteria exhibit almost limitless variety in their metabolism and ability to decompose diverse substrates. Most species of bacteria function as *heterotrophs*, that is, as organisms which require complex organic molecules for growth. Enzymes are secreted which break down the macromolecules into relatively simple, soluble substances (glucose, amino acids) which can be absorbed by the organisms. The remainder function as *autotrophs*, which can synthesize their cell constituents from simple inorganic molecules by harnessing the energy of sunlight (the photosynthetic bacteria), or energy from the chemical oxidation of inorganic compounds (the chemoautotrophs). The bacteria may be further subdivided on the basis of their requirement for molecular oxygen into:

Aerobes—those requiring O_2 as the terminal acceptor of electrons in respiration;

Facultative anaerobes—those normally requiring O_2 but able to adapt to oxygen-free conditions by using NO_3^- and other inorganic compounds as electron acceptors in respiration;

Obligate anaerobes—those which grow only in the absence of O_2.

The activities of some of the specialized groups of bacteria will be discussed in Chapters 8 and 10.

FUNGI

A germinating fungal spore develops a long colourless thread, the *hypha*, which may grow straight or with branching to form a network of hyphae, the *mycelium*. Spores that are often coloured are produced in fruiting bodies on the mycelium. Thus, members of the Basidiomycetes (white-rot fungi) and the abundant spore-formers of the Fungi Imperfecti, such as *Penicillium* and *Aspergillus*, are often conspicuous on decaying wood and leaf litter in moist situations.

Fungi occur as *soil inhabitants*—those residing in the soil and feeding on dead organic matter (the saprophytes), and as *soil invaders*—those whose spores are normally deposited in the soil but which only germinate and grow when living tissue of a suitable host plant appears close by. Many of the latter fungi are pathogenic, growing parasitically on the host. Of special benefit, however, are the *mycorrhizal fungi* which live symbiotically in the host tissue, deriving C compounds from the host and in turn supplying it with mineral nutrients, especially P (section 10.3).

Although soil fungal colonies are much less numerous than bacteria ($1-4 \times 10^5$ organisms g^{-1}), the biomass can be as much as one-half to two-thirds of the bacterial biomass. The fungal population is also considerably less variable than bacterial numbers, for although fungi, excluding yeasts, are intolerant of anaerobic conditions, they grow better in acid soils (particularly at pH < 5.5) and tolerate variations in soil moisture better than bacteria. The fungi are all heterotrophic and are most

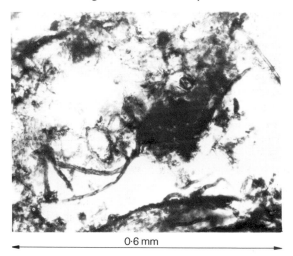

0.6 mm

Figure 3.3. Fungal mycelium in litter layer (courtesy of P. Bullock).

abundant in the litter layer (Figure 3.3) and the organically-rich surface horizons of the soil, where their superior ability to decompose lignin confers a competitive advantage. Because of their slow growth rate relative to the bacteria, fungi cannot be reliably

counted by the dilution plate method; methods employing the transfer of fungal hyphae from the soil to nutrient agar plates give more accurate results.

For details of methods of species identification and counting of soil micro-organisms, the reader is referred to Parkinson, Gray and Williams (1971).

ACTINOMYCETES

These organisms form colonies similar to those of the fungi, with abundant spore formation, but they are slow growing and often difficult to isolate in the presence of bacteria and fungi. Those species which produce antibiotics can be isolated more readily. Aerobic and less tolerant of soil acidity than the fungi, the actinomycetes are better able to decompose lignin and complex organic compounds synthesized in the soil than either fungi or bacteria. Their numbers are comparable to the fungi but their biomass is much less.

ALGAE

The algae are photosynthetic and therefore confined to the soil surface, although they will grow heterotrophically in the absence of light if simple organic solutes are provided. Most important are the blue-green algae, such as the genus *Nostoc*, which can reduce atmospheric nitrogen and incorporate it into amino acids, thereby making a substantial contribution to the N status of the soil (section 10.2). The 'blue-greens' prefer neutral to alkaline soils whereas the green algae are more common in acid soils. Algae in soil are much smaller than the aquatic or marine species and may number from 100,000 to 3×10^6 organisms g^{-1}.

PROTOZOA

The smallest of the soil animals, ranging from 5 to 40 μm in their longer dimension, the protozoa live in water films and move by means of cytoplasmic streaming, cilia or flagellae in the case of *Euglena*. *Euglena* also contains chlorophyll and can live autotrophically, but nearly all of the protozoa prey upon other small organisms such as bacteria, algae and even nematodes. They can be important in regulating bacterial numbers.

The mesofauna

The most important groups from the viewpoint of

organic matter turnover are the arthropods, annelids (earthworms) and molluscs.

ARTHROPODS

This group includes the mites, springtails, beetles and many insect larvae, ants, termites, millipedes and centipedes.

Mites and springtails are the most numerous, feeding on detritus in the lower part of thick litter layers under forest and undisturbed grassland. They may also be found at depth in the soil, living in the larger pores (0.3 to 5 mm). Some species of mite are predatory on springtails.

Many *beetles* and *insect larvae* live in the soil, some feeding saprophytically, while others feed on living tissues and can be serious pests of agricultural crops. Of particular value, however, are the coprophagous beetles of Africa and other tropical regions which feed upon the dung of large herbivores and greatly increase the rate at which such residues are comminuted and physically mixed with the soil (Figure 3.4).

are surface feeders and build nests called termitaria by packing and cementing together soil particles with organic secretions and excrement. The small mounds only 30 cm or so high are the homes of *Cubitermes* species which feed under the cover of recent leaf fall; the infrequent large mounds, some 4–6 m high, are the homes of wood-feeders and foraging termites, such as *Macrotermes* and *Odontotermes*, the latter usually colonizing the mounds of *Macrotermes* when they are abandoned. These species cultivate 'fungus gardens' within the termitaria to provide food for their larvae.

Some termite species or 'harvester ants' destroy living vegetation in African grasslands. In cool temperate regions, species of ant such as *Formica fusca* are of some importance in comminuting litter under pine trees, while *Lasius flavus* is active in grassland soils on chalk.

Centipedes are carnivorous as opposed to *millipedes* which feed on vegetation, much of which is in the form of living roots, bulbs and tubers. Size comparisons between these and other small soil invertebrates may be made by reference to the specimens in Figure 3.5.

Figure 3.4. Dung beetle (*Sisyphus rubripes*) rolling a 'dung ball' (after Waterhouse, 1973).

The *termite* has been called the tropical analogue of the earthworm and indeed they are important comminuters of all forms of litter—tree trunks, branches and leaves—in the forest and especially the seasonal rainfall regions (savanna) of the Tropics. Most species

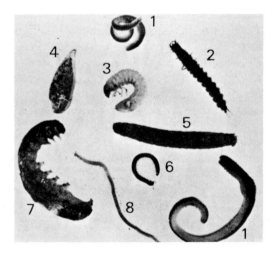

Figure 3.5. The main litter-feeding soil animals (after Russell, 1973). 1. Earthworm, *Lumbricus* sp. (Oligochaeta); 2. Beetle larva, Carabidae (Coleoptera); 3. Chafer grub, *Phyllopertha* sp. (Coleoptera); 4. Slug, *Agriolimax* sp. (Mollusca); 5. Leatherjacket, Tipulidae (Diptera); 6. Millipede (Diplopoda); 7. Cutworm, Agrotidae (Lepidoptera); 8. Centipede (Chilopoda).

EARTHWORMS

On account of their biomass and physical activity, earthworms are generally more important in the consumption of leaf litter than all the other invertebrates together. Exceptions occur in the drier savanna-region soils and soils of temperate regions on which mor humus forms (section 3.3). An indication of the earthworm biomass under different systems of land use is given in Table 3.4.

Table 3.4. Estimates of earthworm biomass in soils under different land use.

	kg ha^{-1}
Hardwood and mixed woodland	370–680
Coniferous forest	50–170
Orchards (grassed)	640–287
Pasture	500–1500
Arable land	16–760

(after Edwards and Lofty, 1977)

Earthworms feed exclusively on dead organic matter. A few species live only in the litter layer; the majority migrate between the litter layer and the mineral soil, and in the course of feeding ingest large quantities of clay and silt-size particles. The large populations of earthworms in soils long down to grass may consume up to 90 t of soil ha^{-1} annually. As a result, the organic and mineral matter is more homogeneously mixed when deposited in the worm faeces, which may appear as *casts* on the soil surface, as for example with *Allolobophora longa* and *A. nocturna*. Other species, such as *A. caliginosa* and *Lumbricus terrestris*, ar enon-casters and deposit their faeces in burrows. Organic matter from the surface rapidly becomes incorporated throughout the upper part of the soil profile, in contrast to the sharp boundary between the litter layer and mineral soil proper in the absence of earthworms (Figure 3.6a and b).

The consumption of dung and vegetable matter by earthworms can be prodigious. It has been estimated from feeding experiments that a population of 120,000 adult worms ha^{-1} is capable of consuming 25–30 t of cow dung annually (Edwards and Lofty, 1977). The quantities of dung and leaf litter available in the field

Figure 3.6a. Litter accumulation on soil devoid of earthworms (after Edwards and Lofty, 1977).

Figure 3.6b. Uniform organic matter distribution in a soil with earthworms (after Edwards and Lofty, 1977).

are normally much less than this, so that the size of the earthworm population is regulated by the amount of suitable organic matter available, as well as by soil temperature, moisture and pH. For example, earthworms are rarely found in soils of pH < 4.5, and most species prefer neutral to calcareous soils. In hot dry weather they burrow deeper into the soil and aestivate.

Grass and most herbaceous leaf litter is readily acceptable to earthworms, but there is great variability in the palatability of litter from temperate forest species. Litter from elm, lime and birch is more palatable than pine needles which must age considerably before being acceptable.

Molluscs. Of the animals in this group (the snails and slugs), many feed on living plants, and are therefore pests, although some species feed on fungi and the faeces of other animals. Since the mollusc biomass is only some 200–300 kg ha^{-1} in most soils, their contribution to the turnover of organic matter is limited.

3.3   CHANGES IN PLANT REMAINS DUE TO THE ACTIVITIES OF SOIL ORGANISMS

Types of humus

Decomposition begins with the invasion of ageing plant tissue by surface saprophytes, and proceeds in parallel with biochemical changes in the senescing tissue—the synthesis of protease enzymes, the rupture of cell membranes with consequent mixing of cellular constituents, and the oxidation and polymerization of phenolic-type compounds. Decomposition accelerates, as evidenced by the rise in the rate of CO_2 production, when the plant material falls to the ground and is invaded by a host of soil organisms. The changes that occur in the plant residues lead not only to mineralization and immobilization of nutrient elements, but also to the synthesis of new compounds, less susceptible to decomposition, which collectively form a dark-brown to black, amorphous material called *humus*.

MOR AND MULL HUMUS

Polyphenolic compounds formed in the senescing leaf which are not leached out by rainwater once it has

fallen to the ground have a considerable effect on the rate of decomposition of the leaf residue. This is well illustrated by the contrasting superficial layers and A horizons of soils under temperate forests.

Typically under deciduous species, there is a loose litter layer 2–5 cm deep under which the soil is well aggregated and porous, dark-brown in colour and changes only gradually with depth to the lighter colour of the mineral matrix. This deep organic A horizon,

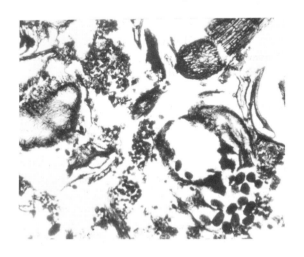

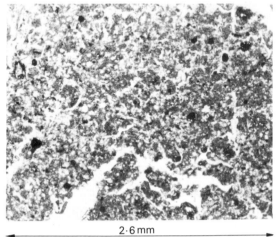

2·6 mm

Figure 3.7a. Loose fabric of plant residues and mite droppings in mor humus (courtesy of P. Bullock).

Figure 3.7b. Intimately mixed matrix of organic and mineral matter in mull humus (courtesy of P. Bullock).

rich in animals especially earthworms, is characteristic of *mull humus*. Alternatively, under coniferous species, the surface litter is thick (5–20 cm) and often ramified by plant roots and a tough fungal mycelium; it is sharply differentiated from the mineral soil which is capped by a thin band of blackish humus and usually compact, poorly drained and devoid of earthworms. This superficial organic horizon is called *mor humus* in which three layers may be recognized:

L layer—undecomposed litter, primarily remains deposited during the previous year.

F layer—the fermentation layer, partly decomposed, with abundant fungal growth and the faecal pellets of mites and springtails.

H layer—the humified layer, comprising completely altered plant residues and faecal pellets.

The contrasting microstructure of mor and mull humus is illustrated by the thin sections of Figure 3.7.

Mor and mull humus, and intermediate types called *mor moder* and *mull moder*, occur under a variety of vegetation types: as well as the main deciduous trees (oak, elm, ash and beech), mull usually occurs under a well drained grassland, whereas mor is found under heath vegetation (*Erica* and *Calluna* spp.) as well as under the major conifers (pine, spruce, larch and fir). However, the tendency towards mor or mull formation is determined not only by plant species, but also by the mineral status of the soil, particularly of Ca, N and P, so that the same species of deciduous tree can form mull on a fertile calcareous soil and mor on an infertile acid sand.

Colonization by soil micro-organisms

The formation of mor humus is predisposed by high concentrations of phenols in the senescing leaf which precipitate cytoplasmic proteins onto the mesophyll cell walls, thereby rendering both the protein and cellulose more resistant to microbial decomposition. The process has been likened to the 'tanning' of leather, whereby plant tannins (polyphenols) are used to preserve the protein of animal skins. Where plant proteins and cellulose are not protected by tanning, these constituents and the sugars and storage carbohydrates are rapidly metabolized by the soil micro-

organisms which colonize the litter in a fairly definite sequence:

mould fungi and non-spore-forming bacteria→ spore formers→cellulose myxobacteria (the slime bacteria)→actinomycetes.

The appearance of the myxobacteria is delayed because these organisms can utilize inorganic N only, so that their development is retarded until the early colonizers have mineralized enough of the organic N.

The role of soil animals

There is a great deal of interdependence between the activities of micro-organisms and the soil mesofauna in the decomposition of litter and formation of humus. The digestive processes of earthworms, mites and springtails are very inefficient, and it seems that some microbial decomposition of the litter must occur before it is readily acceptable to these animals. Once ingested, the main effect of an earthworm's digestive activity, for example, is the comminution of the plant residues which vastly increases the surface area accessible to microbial attack, once the material is excreted. There is some chemical breakdown, usually less than 10 per cent, which could be due to the activities of bacteria in the earthworm gut or to the effect of chitinase and cellulase which are secreted by the earthworms. The combined result of these processes is that earthworm casts have a higher pH, exchangeable Ca^{2+}, available phosphate and mineral nitrogen content than the surrounding soil.

Furthermore, whereas the partially humified residues in the L and F layers of mor humus retain a recognizable plant structure for many years, similar residues which have passed through the earthworm's digestive tract several times are reduced to a black, amorphous humus which is indissociable from the mineral particles.

Biochemical changes and humus formation

Our conception of the biochemical changes involved in humification has changed considerably since Waksman (1938) proposed that lignins condensed with amino acids to form ligno-protein complexes resistant to further decomposition. Kononova (1966) reported that humic materials form in plants before the lignaceous tissues have begun to decompose, and the higher the content of proteins and carbohydrates, as opposed to lignin, the more rapidly are the plant residues humified. It is now thought that the precursors of humic compounds are a variety of simple aromatic compounds, of the kind presented in Figure 3.8, which are derived

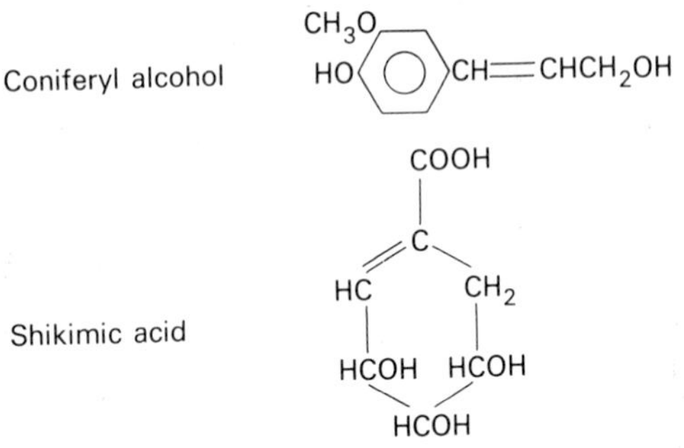

Figure 3.8. Simple aromatic compounds of plant and microbial origin.

either directly from the plant tissues or as the products of microbial metabolism of carbohydrates. These compounds are oxidised, the reaction being catalysed by enzymes secreted by the mould fungi, myxobacteria and actinomycetes, and condense with other oxidation products and amino acids to form humic polymers that contain 2–5 per cent N and a high density of carbonyl (—C=O), carboxyl (—COOH) and phenolic (◎ OH) groups.

The type and number of constituent molecules are legion, as is the number of ways in which they may combine, but spectrophotometric (infra-red, visible and ultra-violet spectra) and spectrometric (electron spin resonance) analyses of the 'humus fractions' (section 3.4) indicate that the major components of humus are fatty acid esters, phenolic acids and benzene polycarboxylic acids (see Figure 3.10).

3.4 PROPERTIES OF SOIL ORGANIC MATTER

Fractionation of organic matter
Physically, soil organic matter may be subdivided into those components which are readily dispersed and

separated from the mineral particles, and that which can only be dispersed by drastic chemical treatment. A suggested subdivision is as follows:

(a) MACRO-ORGANIC MATTER
This consists of the most recently added plant and animal debris which, due to the entrapment of air, has a density near 1 g cm^{-3} and can be separated from the heavier mineral particles by flotation in water or aqueous salt solutions. It is then collected on a fine sieve (0.06 mm aperture).

(b) LIGHT FRACTION
This consists of partially humified and comminuted plant and faunal remains which can comprise up to 25 per cent of the total organic matter in grassland soils. Separation from the mineral particles is achieved after ultrasonic vibration of the dry soil, to disrupt the soil aggregates, and flotation of the light fraction in a heavy organic solvent (density 2 g cm^{-3}) with the aid of a surfactant.

(c) HUMIFIED FRACTION
Much of the soil organic matter adheres strongly to the mineral particles, particularly the clay, to form a *clay–humus complex*. This organic matter cannot be separated by density flotation, and is incompletely dissolved in metal-chelating solvents such as acetylacetone. It is most effectively dispersed by strong alkalis, such as sodium hydroxide, which act both by hydrolyzing and depolymerizing the large organic molecules that are tightly adsorbed on the mineral surfaces.

Traditionally, the humus is subdivided further according to its solubility in alkaline and acidic solutions, as summarized in Figure 3.9. Central to the problem of the chemical characterization of humus is the fact that, because drastic treatment is required to extract the humic material and subsequently to resolve its constituents, it is doubtful that the compounds in the extracts are the same as those originally in the soil. Recent work indicates that macromolecules in the *fulvic acid* (FA) and *humic acid* (HA) fractions, while differing in size, are composed of essentially the same subunits of plant and microbial origin. The partial structure of FA-type material shown in Figure 3.10

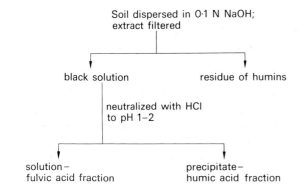

Soil dispersed in 0·1 N NaOH; extract filtered

black solution

residue of humins

neutralized with HCl to pH 1–2

solution – fulvic acid fraction

precipitate – humic acid fraction

Figure 3.9. Scheme for the fractionation of soil humus.

consists of benzene–carboxylic and phenolic acid compounds linked together by relatively weak forces such as hydrogen-bonds and van der Waals forces. The FA fraction also contains polysaccharides primarily of microbial origin, which exist as linear flexible molecules. The higher M.W. (20,000 to 100,000) HA molecules are more stable than the FA molecules probably because of some covalent (C—C and C—O) bonding between subunits. *Humins* are thought to be HA-type

Figure 3.10. Partial structure of fulvic acid (after Schnitzer, 1977).

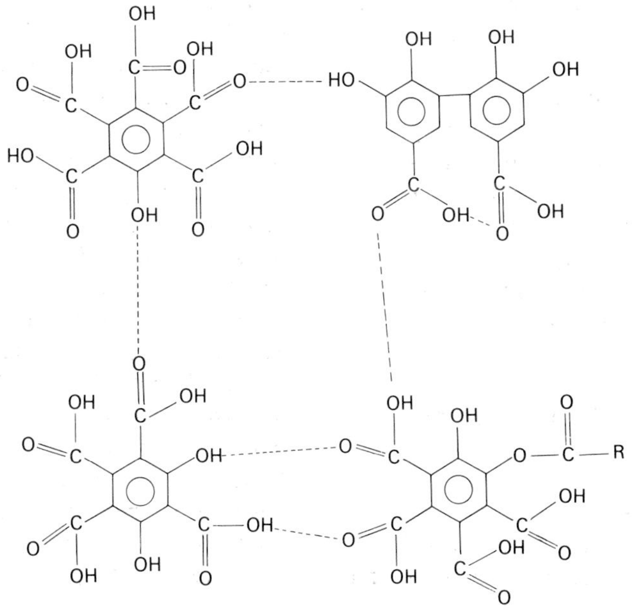

molecules that are very tightly bonded to min surfaces.

CATION EXCHANGE CAPACITY

Humification produces an organic colloid of high specific surface and high cation exchange capacity. Of the several functional groups containing oxygen in the humic compounds, those which dissociate H^+ ions— the carboxylic and phenolic groups—are the most important. The former have pK values ~ 4, and therefore are fully dissociated above pH 7 when the phenolic-OH groups begin to dissociate. Thus, the *CEC* of soil organic matter is completely pH-dependent and buffered over a wide range of H^+ ion concentration; its value ranges between 150 and 300 me per 100 gram dry weight. Organic matter can make a substantial contribution to the *CEC* of the whole soil, and hence to the retention of exchangeable cations, especially in soils of low clay content.

CHELATION

Organic compounds with a suitable spatial arrangement of reactive groups (—OH, ⊚ OH and C꞊O) can form covalent bonds with metallic cations to form *chelate compounds* (Figure 3.11). Chelates formed with

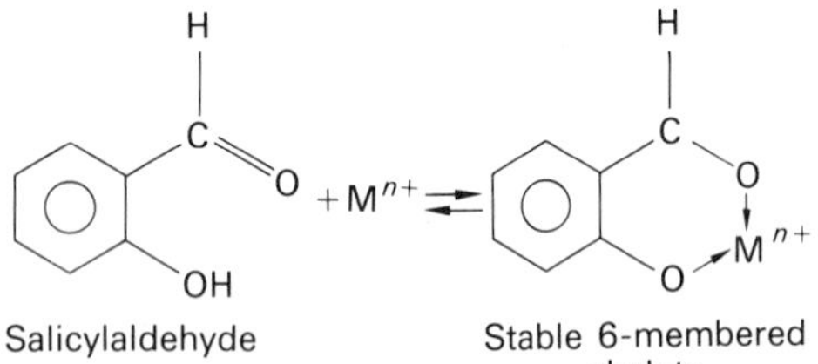

Figure 3.11. Organo-metallic chelate formation.

certain di- and polyvalent cations are the most stable, the stability constants falling in the order $Cu > Fe \simeq Al > Mn \simeq Co > Zn$. Metal complexes formed with the humic acid fraction are largely immobile, but the more ephemeral Fe^{2+}-polyphenolic complexes are soluble and their downward movement contributes to profile differentiation in *podzols* (section 9.2).

Organic matter and soil physical properties
The presence of humified organic matter is of great importance in the formation and stabilization of soil structure (Chapter 4). Low M.W. (< 1000) organic compounds are adsorbed by virtue of their dissociated carboxyl groups at positive sites on sesquioxide films and edge faces of clay minerals, or by displacement of water molecules from the hydration shells of exchangeable cations to form 'cation bridges' (Figure 3.12). The

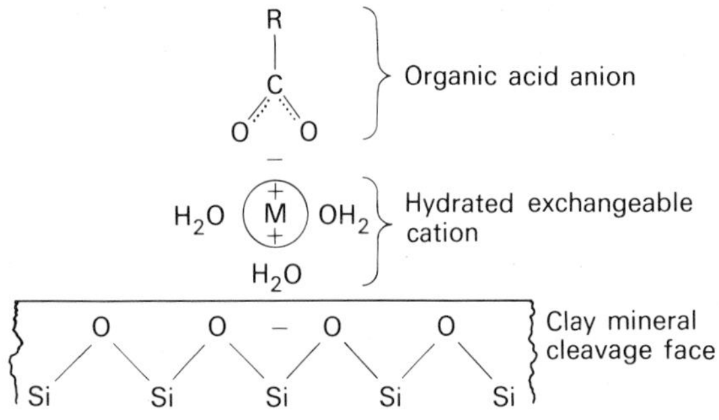

Figure 3.12. Organic acid molecule forming a cation-bridge (after Greenland, 1971).

large uncharged polymers, such as the polysaccharide and polyuronide gums synthesized by many bacteria,

can be adsorbed at mineral surfaces by hydrogen-bonding and by van der Waals' forces (see Table 4.2), and also function as bonding agents between mineral particles.

3.5 FACTORS AFFECTING THE RATE OF ORGANIC MATTER DECOMPOSITION

Turnover
Gains of organic matter through litter deposition and root death are offset by losses through decomposition and leaching. This is called *turnover*, which is defined as the flux of organic C through a unit volume of soil. Given a constant environment, a soil will eventually attain a steady-state equilibrium in which there is no measurable change in the organic matter content with time. The concept of turnover may be expressed mathematically by the equation:

$$\frac{dC}{dt} = A - kC \qquad (3.1)$$

where C = the organic C content of the soil (t ha^{-1}); A = the annual addition of residues, assumed to be constant; and k is a rate constant, assuming decomposition is a simple first-order reaction. At equilibrium, $dC/dt = 0$ and

$$A = kC \qquad (3.2)$$

The *turnover time* is then given by

$$\frac{C}{A} = \frac{1}{k} \text{ (years)} \qquad (3.3)$$

Annual inputs of C for old arable soils at Rothamsted in England have been estimated at 5 to 6 per cent of the soil organic C, giving turnover times of 16 to 22 years. Theoretically, this is the time taken for all the organic C in the soil to be replaced, and in a sense is inversely related to the 'biological activity' of the soil. But the picture is over-simplified because the half-life of C in these soils has been determined from radioactive carbon ^{14}C dating as 1450 to 3700 years, depending on the depth in the soil. It is obvious that there must be a great range of decomposability of C compounds in the soil, reflecting a continuous spectrum of k values.

Jenkinson and Rayner (1977) have successfully modelled the changes in organic C in these Rothamsted soils on the basis of five fractions of different decomposability within the organic matter: they assumed that over 80 per cent of the residues added annually decomposed rapidly, with a constant proportion of the C being released as CO_2 or transferred to the biomass and pools of physically and chemically stabilized organic matter. The balance remained as resistant plant residues. These transformations, with the appropriate rate constants, are represented in Figure 3.13. The model predicted that, for an annual

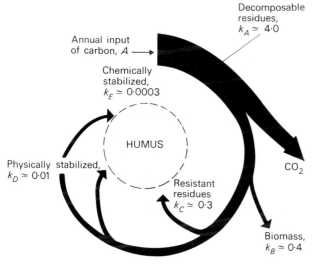

Figure 3.13. Model for the transformations of an annual input of carbon in the soil (after Jenkinson and Rayner, 1977).

input of 1 t C ha^{-1}, the soil at equilibrium would have more than half its C in a very resistant form, the half-life of which was 1980 years.

Substrate availability

PRIMING ACTION
It was once thought that the stimulus to the resident soil microbial population provided by the addition of fresh residues was often sufficient to accelerate also the decomposition of soil humus. This effect was called *priming action*. Priming action, which can be positive or negative, has been demonstrated subsequently by

following the decomposition of ^{14}C-labelled residues, as illustrated in Figure 3.14. The conclusion from this

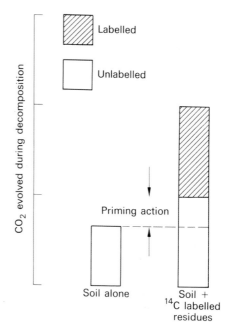

Figure 3.14. Priming action measured by the use of ^{14}C-labelled residues (after Jenkinson, 1966).

kind of experiment is that priming action generally has a negligible effect on the overall rate of decomposition. The addition of fresh residues does stimulate microbial activity, but the increase in soil biomass usually compensates for any acceleration in the rate of humus decomposition, so that the net effect is an increase in soil organic matter.

CLAY CONTENT AND CULTIVATION
The adsorption of protein by montmorillonite, especially in the interlamellar regions, protects the protein from microbial attack. Similarly, the adsorption of various compounds by clays and sesquioxides generally serves to slow down their rate of decomposition. The organic matter held in the relatively stable pores in clay soils of diameter <1 μm—the so-called 'sterile pores'—is also less accessible to microbial attack. Cultivation of the soil tends to break down the structure

so that organic matter in sterile pores is more exposed to micro-organisms. Higher surface soil temperatures are also attained when the protection of the vegetative canopy and litter layer is lost.

Under English conditions, clay content has little effect on the rate of decomposition of most fresh residues. Approximately one-third of the added carbon remains after 1 year, and one-fifth after 5 years. The decomposition rate is slightly lower in clay than in sandy soils, but part of this difference is explicable by the greater leaching losses from light-textured soils. Under tropical conditions, where the potential for decomposition is greater, the part played by clay minerals and particularly sesquioxides in protecting soil organic matter is much greater.

Activity of the soil biomass

NUMBERS OF ORGANISMS

The effect of environmental and soil factors on the size and species diversity of the populations of soil micro-organisms has been referred to briefly. Soil animals are also affected by types of litter, prevailing conditions of temperature, moisture and pH, and management practices such as cultivation and the use of agricultural chemicals.

PHYSIOLOGICAL ACTIVITY

Environmental and management factors can influence the physiological activity of the soil organisms. Striking effects occur, for example, when the soil is treated with toxic chemicals, such as chloroform or toluene, which can cause partial to total sterilization. On the restoration of favourable conditions, there is an explosion in microbial numbers resulting in a flush of decomposition as the surviving organisms feed upon the residues of the dead organisms and multiply rapidly. Similar flushes of decomposition occur in soils subjected to extremes of wetting and drying (section 10.2) and in soils of temperate regions on passing from a frozen to thawed state.

The dynamics of soil biomass changes in response to variations in soil and environmental factors are summarized graphically in Figure 3.15.

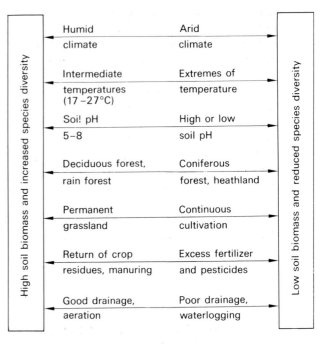

Figure 3.15. Dynamics of soil biomass changes.

3.6 SUMMARY

Plant litter, dead roots, animal remains and excreta form the *soil organic matter* which is a repository for the essential elements required by the next generation of organisms. Release of these elements or *mineralization* depends on the decomposition rate and the demands made by the heterogeneous population of soil organisms.

The living organisms (*biomass*) vary in size from the *macrofauna*—vertebrate animals of the burrowing type, to the *mesofauna*—invertebrates such as mites, springtails, insects, earthworms and nematodes, to the *micro-organisms*, which comprise bacteria, fungi, actinomycetes, algae and protozoa. Earthworm biomass under undisturbed grassland ranges from 0.5 to nearly 3 t ha^{-1}, while the microbial biomass makes up 2–3 per cent of the total organic C in the soil, or 0.5–2 t C ha^{-1}.

The interaction of micro-organisms and mesofauna in decomposing litter leads not only to the release of mineral nutrients but also to the synthesis of complex new organic compounds that are more resistant to attack. This is called *humification*. Annual rates of litter return vary from 0.1 t ha⁻¹ in Alpine and Arctic forests to more than 10 t ha⁻¹ in tropical rain forests. High C:N ratios (>25) favour the increase of microbial biomass and the *immobilization* of N. Low temperature and pH retard decomposition, as does a high polyphenolic content in the senescing plant tissue, leading typically to the formation of *mor humus*. *Mull humus* forms on base-rich soils under herbaceous and deciduous forest species low in polyphenols.

Soil humus has a high *CEC* and is active in chelating metallic cations. Humic compounds and microbial polysaccharides are also important in the stabilization of soil structure (Chapter 4).

REFERENCES

EDWARDS C.A. & LOFTY J.R. (1977) *Biology of Earthworms*, 2nd Ed. Chapman and Hall, London.

JENKINSON D.S. & RAYNER J.H. (1977) The turnover of soil organic matter in some of the Rothamsted classical experiments. *Soil Science* **123**, 298–305.
KONONOVA M.M. (1966) *Soil Organic Matter*, 2nd Ed. Pergamon, Oxford.
PARKINSON D., GRAY T.R.G. & WILLIAMS S.T. (1971) *Ecology of Soil Micro-organisms*. IBP Handbook No. 19, Blackwell Scientific Publications, Oxford.
WAKSMAN S.A. (1938) *Humus*, 2nd Ed. Bailliere, Tindall and Cox, London.

FURTHER READING

BURGES A. (1958) *Micro-organisms in the Soil*. Hutchinson, London.
HANDLEY W.R.C. (1954) Mull and mor formation in relation to forest soils. *Forestry Commission Bulletin 23*. HMSO, London.
JACKSON R.M. & RAW F. (1966) *Life in the Soil*. Edward Arnold, London.
RUSSELL E.J. (1957) *The World of Soil*. Collins, London.
SCHNITZER M. (1977) Recent findings on the characterization of humic substances extracted from soils from widely differing climatic zones, in *Soil Organic Matter Studies. Proceedings IAEA/FAO Symposium*, Braunschweig, 117–131.

Chapter 4
Peds and Pores

4.1 SOIL STRUCTURE

As we have seen in Chapter 3, one reason why the soil supports a diversity of plant and animal life is the abundance of energy and nutrients supplied in organic remains. An equally vital reason lies in the physical protection and favourable temperature, moisture and oxygen supply afforded by the structural organization of the soil particles. Weathering of parent material produces the primary soil particles of various sizes—the clay, silt, sand and stones. These particles may simply 'pack', as for example in Figure 4.1a, to attain a state of minimum potential energy. However, vital forces associated with plants, animals and micro-organisms and physical forces associated with the change in state of water and its movement, arrange the soil particles into a definite pattern or structure (Figure 4.1b). Several definitions of *soil structure* have been

Figure 4.1. (a) Random close packing of soil particles. **(b)** Structured arrangement of soil particles.

(a)

(b)

offered: in the broadest sense, it denotes

(a) the size, shape and arrangement of the particles and aggregates;

(b) the size, shape and arrangement of the voids or spaces separating the particles and aggregates;

(c) the combination of voids and aggregates into various types of structure.

4.2 LEVELS OF STRUCTURAL ORGANIZATION

Many soils have structural features that are readily observed in the field, and the most widely used system for describing and classifying these features—the *macrostructure*—is that laid down in the USDA Soil Survey Manual of 1951. However, much of the fine detail of aggregation and the distribution of voids—the *microstructure*—can only be determined in the laboratory using powerful tools such as the polarizing light microscope, and the scanning electron microscope which can provide magnification up to 20,000 times. Some of the macroscopic features have identifiable counterparts at the microscopic level, in which case a descriptive term common to both macro- and microstructure can be used; other features are peculiar to the microstructure and have therefore required new descriptive terms which have been devised by specialists in the study of thin soil sections—notably Kubiena (1938) and Brewer (1964).

Aggregation

Relatively permanent aggregates or *peds* are recognizable in the field soil because they are separated by voids and natural planes of weakness. They should persist through cycles of wetting and drying, as distinct from the less permanent aggregates formed at or near the soil surface by the mechanical disturbance of digging and ploughing (*clods*) or the rupture of the soil mass *across* natural planes of weakness (*fragments*). A *concretion* is formed when localized accumulations of an insoluble compound irreversibly cement or enclose the soil particles. *Nodules* are similarly formed but lack the symmetry and concentric internal structure of concretions.

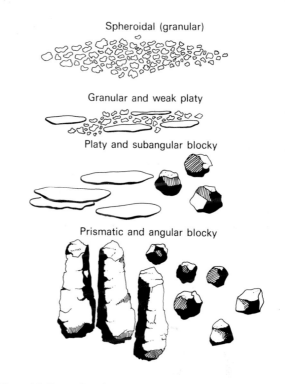

Figure 4.2. Examples of common ped types (after *Modern Farming and the Soil*, 1970).

PED TYPE

Four main types of ped have been recognized, illustrations of which are provided in Figure 4.2. A brief description of each type follows:

(1) *Spheroidal*. Peds that are roughly equidimensional and bounded by curved or irregular planar surfaces that are not accommodated by the faces of adjacent peds. There are two subtypes—*granules* which are relatively non-porous, and *crumbs* which are very porous. These peds are usually evenly coloured throughout, and arise primarily from the interaction of soil organisms, roots and mineral particles. They are therefore common in the A horizons of soils under grassland and deciduous forest.

(2) *Blocky*. Peds that are bounded by curved or planar surfaces that are more or less mirror images of the faces of surrounding peds. Those with flattened

Figure 4.3. Prismatic structure of a clay subsoil prone to water-logging.

faces and sharply angular vertices are called *angular*; those with a mixture of rounded and flat faces and more subdued vertices are called *subangular*. Blocky peds are features of B horizons where their natural surfaces may be identified by their smoothness and often distinctive colours. The ped faces may be coated, for example, by organic matter, clay or mineral oxides (section 4.3). The voids between peds often form paths for root penetration and the ped faces may have live or dead roots attached, or show distinct root impressions.

(3) *Platy*. The particles are arranged about a horizontal plane with limited vertical development; most of the ped faces are horizontal. When the ped is thicker in the middle than at the edges it is called *lenticular*. These peds may occur in the A horizon of a soil immediately above an impermeable B horizon, or in the surface of a soil compacted by heavy machinery.

(4) *Prismatic*. In these peds the horizontal development is limited compared to the vertical; the peds are bounded by flat, vertical faces with sharp vertices. They are subdivided into prisms without caps (prismatic) and prisms with rounded caps (*columnar*). Prismatic structure is common in the subsoil of heavy clay soils subject to frequent waterlogging (Figure 4.3). Columnar structure is associated particularly with the B horizon of sodium-affected soils (Figure 4.4).

COMPOUND PEDS
All types of ped shown in Figure 4.2 are invariably compounded of smaller units which are clearly visible. and others which can only be seen under a microscope. For example, the large compound prismatic peds in Figure 4.5 are composed of secondary subangular blocks which in turn break down to the primary structure of angular blocky peds < 1 cm in diameter. Each of these primary peds presents an intricate internal arrangement of granules and voids which can only be elucidated by careful appraisal of the soil in thin sections (section 4.3).

CLASS AND GRADE
In addition to type, the *class* of ped is judged according to its size, and the *grade* according to the distinctness and durability of the visible peds. Grade of structure may be described as follows:

(1) *Structureless (or apedal)*. No observable aggregation nor definite arrangement of natural lines of weakness. *Massive* if coherent; *single grain* if non-coherent.

(2) *Weakly developed*. Poorly formed, indistinct peds that are barely observable *in situ*. When disturbed, the soil breaks into a mixture of a few entire peds, many broken peds and much unaggregated material.

(3) *Moderately developed*. Well formed, distinct peds

Figure 4.4. Columnar structure in the B horizon of a *solodized solonetz* (CSIRO photo, courtesy of G.D. Hubble).

Figure 4.5. Compound prismatic peds showing the breakdown into secondary and primary structure.

Prismatic compound ped

Subangular blocky secondary ped

5 cm

that are moderately durable, but not sharply distinct in undisturbed soil. The soil, when disturbed, breaks pown into a mixture of many entire peds, some broken deds and a little aggregated material.

(4) *Strongly developed.* Durable peds that are quite evident in undisturbed soil; they adhere weakly to one another and become separated when the soil is disturbed. The soil material consists very largely of entire peds and includes a few broken peds and little or no unaggregated material.

Grades of structure reflect differences in the strength of *intraped* cohesion relative to *interped* adhesion. The grade will depend upon the moisture content at the time of sampling and the length of time the soil has been exposed to direct drying in air. Prolonged drying will markedly increase the structure grade, particularly of clay soils and soils of high iron oxide content.

0·6 mm

Figure 4.6. Thin section through a soil pore showing clay coatings (courtesy of P. Bullock).

4.3 SOIL MICROMORPHOLOGY

Techniques and terminology

Thin soil sections and their study come under the heading of *soil micromorphology*. Briefly, the method of thin sectioning consists of taking an undisturbed soil sample (dimensions $10 \times 5 \times 3.5$ cm), removing the water in such a way as to minimize soil shrinkage and impregnating the voids with a resin under vacuum. Once the resin has cured, the soil block is sawn and polished to produce a section ~ 25 μm in thickness which is stuck onto a glass slide for viewing. The section is examined with a petrological microscope to determine the characteristic optical properties of the minerals present in the soil. An example of a thin soil section is seen in Figure 4.6.

To date, no unambiguous set of terms of wide acceptance to the micromorphologists has been agreed.

A summary of the essential descriptive terms defined by Brewer (1964) is presented in Table 4.1.

Reorientation of plasma in situ

During the repeated cycles of wetting and drying, and sometimes freezing and thawing that accompany soil formation, the plate-like clay particles gradually become re-organized from a randomly arranged network of small bundles—the *asepic* S-matrix, into larger bundles which show a degree of preferred orientation—the *sepic* S-matrix. The bundles or plasma separations are called *domains*, which can be up to 10 μm long and 1 to 5 μm wide, depending on the dominant clay mineral present and their mode of formation.

As well as the reorganization of clay particles within peds, there is often a change in the orientation of clay particles at ped faces. In dry seasons, material from upper horizons sometimes falls into deep cracks in clay soils. On rewetting, the subsoil, which now contains more material than before, swells to a larger volume and exerts a pressure that displaces material upwards. Ped faces slide against each other to produce zones of preferred orientation called *slickensides* lying at an angle of about 45° to the vertical (Figure 4.7).

Plasma translocation and concentration

One of the major processes of soil formation is the translocation of colloidal and soluble material from upper horizons into the subsoil. In the changed physical and chemical environment of the horizon of illuviation, salts and oxides precipitate, clay particles flocculate and the ped faces become coated with deposited materials. Brewer has classified these *coatings* or *cutans* according to the nature of the surface to which they adhere and their mineralogical composition. Some of the more common examples are given below.

Clay coatings (argillans). Often different in colour to and with a higher reflectance than the S-matrix of the ped, they are easily recognized in sandy and loam soils, but difficult to distinguish from slickenside surfaces in clay soils (Figure 4.6).

Sand or silt coatings (skeletans). They can be formed

Table 4.1. Summary of descriptive terms used in soil micromorphology.

FABRIC—the spatial arrangement of particles, primary and compound (aggregated), and voids

Soil material with basic units or organization—the PEDS

Soil material without obvious organization—APEDAL

S-MATRIX—soil material (< 20 μm) comprising the plasma, skeleton grains and voids (designated S-matrix by Brewer to distinguish it from the fine-grained matrix so named for sedimentary rocks). Thus we have:

PLASMA
colloidal (< 2 μm) and soluble materials not bound up in skeleton grains, which can be moved, reorganized and/or concentrated during soil formation.

SKELETON GRAINS
larger than colloidal size and include detrital mineral fragments, secondary crystalline and amorphous bodies which are relatively stable and not usually translocated, concentrated or reorganized during soil formation.

VOIDS
the spaces within the plasma and between plasma and skeleton grains.

PEDOLOGICAL FEATURES—those formed due to reorganisation of matrix materials by pedogenic processes (ORTHIC) and those inherited from the parent material (INHERITED). Orthic features are subdivided into:

PLASMA CONCENTRATIONS
the concentration of any component of the plasma which has moved in solution or in suspension.

PLASMA SEPARATIONS
the changed arrangement of the plasma components, as, for example, the orientation of clay minerals at the surfaces of peds, such as slickensides (see text).

FOSSIL FORMATIONS
preserved root channels or burrows of the soil fauna.

(after Brewer, 1964)

Figure 4.7. Intersecting slickensides in a heavy clay subsoil (CSIRO photo, courtesy of G.D. Hubble).

by the illuviation of sand or silt, or by the residual concentration of these particles after clay removal.

Sesquioxidic coatings (sesquans, mangans, ferrans). They can be formed by reduction and solution of Fe and Mn under anaerobic conditions and their subsequent oxidation and deposition in aerobic zones (Figure 4.8). Colours range from black (mangans) through grey, reddish, yellow, blue and blue-green (ferrans), depending on the degree of oxidation and hydration of the iron compounds.

Salt coatings (of gypsum, calcite, sodium chloride) and *organic coatings* (organans) can also occur.

Mixtures of these materials, particularly clay, sesquioxides and organic matter to form *compound coatings* are not uncommon.

Pans. The coatings may build up to such an extent that bridges of material form between soil particles and the smallest peds so that a cemented layer, sometimes impenetrable to roots, may develop in the soil. Such a layer or horizon is called a *pan* and is described by its thickness (if < 1 cm it is called 'thin') and its main chemical constituent. An example of a thin iron pan is shown in Figure 4.9. Compacted horizons which are weakly cemented by silica are called *duripans*. On the other hand a *fragipan* is not cemented at all, but

Figure 4.8. Manganese dioxide coatings on ped faces (courtesy of R. Brewer).

Figure 4.9. A thin iron pan developed in ferruginous sand deposits.

the size distribution of the soil particles is such that they pack to a very high bulk density (section 4.5).

Voids. Cylindrical and spherical voids, usually old root channels or earthworm burrows, are called *pores* and those of roughly planar shape are called *fissures*. Fissures start as cracks striking vertically downwards from the surface, especially in soils containing much of the expanding-lattice type clays (Figure 4.10). Secondary cracks develop at right angles to the vertical fissures.

4.4 THE CREATION OF SOIL STRUCTURE

Origin of the forces

Structure formation requires work to be done on disorganized soil material by physical, chemical and biotic

forces to create peds and pedological features (Table 4.1). Swelling pressures are generated in many clay soils due to the osmotic effects of the exchangeable cations adsorbed by the clay colloids. In addition, there are potent forces associated with the change in state of the soil water. Unlike any other liquid, water expands on cooling from 4° to 0°C. As the water in the largest pores freezes, liquid water is drawn from the finer pores thereby subjecting them to shrinkage forces; but the build-up of ice lenses in the large pores, and the expansion of the water on freezing, subjects these regions to intense disruptive forces. At higher temperatures tension forces develop as water evaporates

Figure 4.10. Close cracking pattern in the surface of a montmoril-lonitic clay soil (scale in photo is 15 cm long) (CSIRO photo, courtesy of G.D. Hubble).

from soil pores, and the prevalence of evaporation over precipitation can lead to localized accumulations of salts and the formation of concretions anywhere within the soil profile. The mechanical impact of rain, animal treading and earth-moving machinery can re-organize soil materials and the energy of running water can move them.

Biotic forces are no less important. Burrowing animals reorganize soil materials, particularly the earthworm which grinds and mixes organic matter with clay and silt particles in its gut, adding extra calcium in the process, to form a cast of considerable strength. The physical binding of soil particles by fine roots and fungal mycelia ramifying through the soil also contri-butes to aggregation, and to the stabilization of aggregates as is discussed below. Further, the metabolic energy expended by soil micro-organisms can maintain concentration gradients for the diffusion of salts, or induce a change in valency leading to the solution and subsequent reprecipitation of manganese and ferric oxides.

The various interparticle forces of attraction which contribute to structure formation are summarized in Table 4.2.

Table 4.2. Summary of interparticle forces of attraction.

ELECTROSTATIC OR COULOMBIC FORCES—these involve clay–clay, clay–oxide and clay–organic interactions and include:

(a) The positively-charged edge of a clay mineral attracted to the negative cleavage face of an adjacent clay mineral.

(b) A positively-charged sesquioxide film 'sandwiched' between two negative clay mineral surfaces.

(c) Positively-charged groups (e.g. amino groups) of an organic molecule attracted to a negative clay mineral surface. Organic molecules, in displacing the more highly hydrated inorganic cations from the clay, considerably modify its swelling properties.

(d) Clay-polyvalent cation—organic anion linkages or 'cation-bridges'. If the organic anion is very large (a polyanion), it may interact with several clay particles through cation-bridges, positively-charged edge faces or isolated sesquioxide films. Linear polyanions of this kind (e.g. 'Krilium') are excellent flocculants of clay minerals (section 7.5).

VAN DER WAALS' FORCES—these involve clay–organic inter-actions such as:

(a) Specific dipole-dipole attraction between constituent groups of an uncharged organic molecule and a clay mineral surface, as in hydrogen-bonding between a polyvinyl alcohol and O or OH surface.

(b) Non-specific dispersion forces between large molecules when in very close proximity, which are proportional to the number of atoms in each molecule. These forces can account for the strong adsorption of polysaccharide and polyuronide gums and mucilages by soil particles. These polymers of molecular weight 100,000–400,000 are produced in the soil by bacteria, such as *Pseudomonas* spp., which are abundant in the rhizosphere of many grasses. When the adsorption of such a large molecule results in the displacement of water molecules, then the free energy decrease on adsorption can be very large and the reaction is essentially irreversible. Large, flexible polymers of this kind can make contact with several clay particles at many points so that they help to hold the particles in an aggregate in the manner of a 'coat of paint' (section 7.5).

Aggregate formation and stabilization

FORMATION

Several ideas are current as to the way in which the constituent particles—sand and silt grains, clay and organic colloids—are organized to form a ped.

One hypothesis is that the stable subunits are microaggregates not greater than 0.25 mm diameter which are formed by the regular packing of composite particles of [clay—polyvalent cations—organic matter]. These microaggregates possess a considerable degree of water-stability, but the forces of attraction between the clay microaggregates and sand grains are weak. An alternative hypothesis proposes that the organic matter is 'wrapped around' the stable subunits of the microstructure—the *clay domains* (section 4.3). The domains may be mutually attracted (positive edges to negative cleavage faces) or bonded to much larger sand grains by organic polymers, in the manner illustrated in Figure 4.11. The stability of the resultant aggregate,

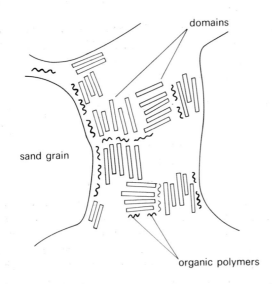

Figure 4.11. Possible arrangements of clay domains, organic polymers and sand grains in aggregate microstructure (after Emerson, 1959).

which is dependent on the number of domain-quartz bonds, is generally much greater under forest and grassland than in arable soils.

Each domain consists of a group of Ca-saturated clay crystals or quasi-crystals, up to 5 μm thick overall, which acts as a single entity when immersed in water. Water penetrates between the clay crystals causing intercrystalline or domain swelling, but not between the individual subunits of each crystal (intracrystalline), which are strongly bonded at an interlamellar spacing of 1 nm or less. The clay crystals can only be disrupted by applying an external force, as with an ultrasonic vibrator, or by replacing much of the exchangeable Ca^{2+} with Na^+. The swelling properties of clays and clay domains are discussed further in section 7.4.

STABILIZATION

As indicated in the discussion on pans, inorganic compounds can act as interparticle cements. For example, calcium carbonate deposits restrict the swelling of the clay colloids in soils formed on calcareous marine sediments (the marls), and calcium carbonate also plays a key role in the stabilization of the framework of silt-size grains in loess soils. Indurated layers cemented by silica can occur at depth in highly weathered *laterites*, while in the upper horizons the residual products of weathering—the iron and aluminium oxides—form bridges between particles of all sizes to form a very stable structure.

However, for soils of pH between 5.5 and 7, which derive little benefit from aggregate stabilization by either sesquioxides or calcium carbonate, organic materials appear to be the most important bonding agents (section 3.4). In some soils, the hydrophobicity of the organic matter 'waterproofs' the mineral surfaces and protects the structure by slowing down the rate of rewetting when the soil is very dry. It has also been discovered that some soil-dwelling Basidiomycetes, such as *Trichoderma viride* and *Rhizoctania solani*, are important in stabilizing aggregates in soil under grassland. These fungi flourish in the undisturbed conditions of the grassland but do not compete well with other soil micro-organisms, so that when the sward is broken up by cultivation, the stability of the structure deteriorates quite rapidly.

The relation between soil structure and tillage, and the influence of land use systems on the maintenance of a good structure is discussed in Chapter 11.

4.5 PORE VOLUME

Porosity

Clearly, the size, shape and arrangement of the peds determines the pore space or *porosity* of the soil. Porosity is measured as the ratio:

$$\text{Pore Space Ratio } (PSR) = \frac{\text{volume of pores}}{\text{total soil volume}} \quad (4.1)$$

The *PSR* depends on the water content of the soil, since both pore volume and the total volume of an initially dry soil may change due to the swelling pressures developed as clay surfaces hydrate within the peds. The soil moisture content at which the porosity is measured therefore needs to be stated.

Total pore space tells us nothing about the actual size distribution of the pores. Pores larger than about 60 μm in diameter can be seen with the naked eye, and are called *macropores*. They comprise a high proportion of the space, particularly that created by faunal activity, between the soil peds. The distribution of pore sizes in a soil, especially a non-swelling one, can be determined reasonably accurately from the volume of water released in response to known suctions, using the relationship between suction and the radius of pores that will just retain water at that suction (section 6.1).

Bulk density

Porosity can be measured from the volume of a non-polar liquid, such as paraffin, which is absorbed into a dry ped under vacuum. Alternatively, porosity may be determined indirectly from the bulk density and particle density. Particle density (ρ_p) was introduced in equation 2.1 (p. 11); *bulk density* (*B.D.*) is defined as the weight of oven-dry soil per unit volume, and depends on the densities of the constituent soil particles (clay, organic matter etc.) and their packing arrangement. Thus, for a sample of oven-dry (o.d.) soil, from equation 4.1:

$$PSR = \frac{\text{total soil volume} - \text{volume of solids}}{\text{total soil volume}}$$

$$= 1 - \frac{\text{weight of o.d. soil}}{\text{total soil volume}} \times \frac{\text{volume of solids}}{\text{weight of o.d. soil}}$$

$$\text{i.e. } PSR = 1 - \frac{\text{bulk density } (B.D.)}{\text{particle density } (\rho_p)} \quad (4.2)$$

Values of *B.D.* range from <1 g cm^{-3} for soils high in organic matter, to 1.0–1.4 for well aggregated loamy soils, to 1.2–2.0 for sands and compacted horizons in clay soils. Using equation 4.2, it can easily be calculated that a sandy soil of bulk density 1.50 and particle density 2.65 has a *PSR* value of 0.43 or 43 per cent. Soil measurements made on a unit weight basis are translated to a unit volume basis by multiplying by the *B.D.*; where the exact *B.D.* is not known, it is common practice to assume an 'average' value of 1.33, which corresponds to a soil weight of 2.0×10^6 kg ha^{-1} to a depth of 15 cm.

Water-filled porosity

The pores of a soil are partly or wholly occupied by water. Water is vital to life in the soil, its importance stemming from its existence as a liquid over the temperature range most suited to living organisms. It possesses several unique physical properties, namely the greatest specific heat capacity, latent heat of vaporization, surface tension and permittivity* of any known liquid.

Soil water is arbitrarily defined as the water which is lost by heating the soil for at least 24 hours to a temperature of 105°C. The moisture content is expressed as a percentage by weight, i.e.

$$\text{H}_2\text{O content } (\%) = \frac{\text{weight of water}}{\text{weight of oven-dry soil}} \times 100$$

$$(4.3)$$

or as a fraction by volume:

$$\theta = \frac{\text{volume of water}}{\text{volume of dry soil}} \quad (4.4)$$

Structural water, that is water of crystallization held within the soil minerals, is not driven off at a temperature of 105°C so that it is excluded from the definition of soil water. Similarly, a distinction is made between soil water and *groundwater* which occurs at a variable depth below the soil surface where the soil and parent material are permanently saturated. The top level of the groundwater is called the *watertable*.

Normally water occupies less than the total pore space and the soil is said to be unsaturated, although after a period of heavy rain, the soil may temporarily

* Dielectric constant.

be saturated to the surface. The volumetric water content θ gives a direct measure of the equivalent depth of water in the soil, which is useful for comparison with the rainfall, evaporation or the amount of irrigation water applied. An 'average' value for θ of 0.25 translates into 25 cm of water per metre depth of soil.

Air-filled porosity

Those pores and fissures which do not contain water are filled with air. The air-filled porosity is defined by:

$$\epsilon = \frac{\text{volume of soil air}}{\text{volume of dry soil}} \quad (4.5)$$

and it follows that

$$\epsilon = PSR - \theta \quad (4.6)$$

Soils have been found to drain, after rain, to a water content in equilibrium with suctions between 50 and 100 mbars. This water content defines the field capacity, of which more will be said in Chapter 6. For many well structured British soils, the field capacity is attained at suctions close to 50 mbars when all the pores $> 60\ \mu m$

diameter are drained of water. The value of ϵ for a soil in this state defines the *air capacity*, C_a (50), which as Figure 4.12 shows, may vary from 0.1 to 0.30 for topsoil, depending on the texture and structure. A knowledge of the air capacity is vital for the planning of agricultural under-drainage schemes (Chapter 13).

Composition of the soil air

The major gases in the Earth's atmosphere normally occur in the proportions 79% N_2, 21% O_2 and 0.03% CO_2 by volume. According to Dalton's law of partial pressures, these concentrations correspond to partial pressures of 0.79, 0.21 and 0.0003 of one atmosphere, which, are equivalent to pressures of 0.802, 0.213 and 0.0003 bars, respectively.

The composition of the soil air may deviate from these proportions owing to soil respiration (section 8.1) which depletes the air of O_2 and raises the content of CO_2. The exchange of O_2 and CO_2 between the atmosphere and the soil air tends to establish an equilibrium which depends on the soil respiration rate and on the resistance to gas movement through the soil. Thus, for a soil in which the gases diffuse rapidly through the macropores, O_2 partial pressures are ~ 0.20 bars and CO_2 pressures ~ 0.005 bars: the normal gas compositions of a variety of soils put to different uses are given in Table 4.3.

Table 4.3. The composition of the gas phase of well aerated soils.

Soil and land use	Usual composition	
	O_2	CO_2
	(% by volume)	
Arable land, fallow	20.7	0.1
unmanured	20.4	0.2
manured	20.3	0.4
Sandy soil, manured and cropped with		
potatoes	20.3	0.6
serradella	20.7	0.2
Pasture land	18–20	0.5–1.5

(after Russell, 1973)

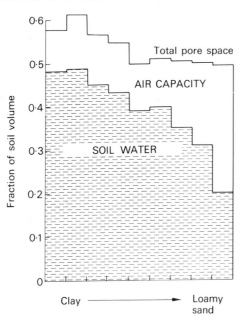

Figure 4.12. Air and water-filled porosities at field capacity in top-soils of varying texture (SSEW Technical Monograph No. 9, 1977).

However, if the macropores begin to fill with water and the soil becomes waterlogged, undesirable chemical and biological changes ensue as the soil becomes increasingly O_2 deficient or *anaerobic*; the O_2 partial pressure may drop to zero and that of CO_2 rises above 0.10 bars. In addition, micro-organisms adapted to living under anaerobic conditions proliferate and produce, apart from CO_2, many unusual waste products and reduced forms of N, Mn, Fe and S. These effects are discussed more fully in Chapter 8.

4.6 SUMMARY

Physical, chemical and biotic forces act upon the primary clay, silt and sand particles, in intimate combination with organic materials, to form an organized arrangement or *soil structure*. On a microscopic scale, clay crystals in roughly parallel array form *domains*, up to 5 μm thick, which are reasonably stable in water provided that Ca^{2+} is the dominant exchanageable cation. Mutual attraction of positively and negatively charged surfaces of clay minerals, sesquioxides and organic polyanions, reinforced by van der Waals forces and hydrogen-bonding, results in the build up of microaggregates. The cohesion of microaggregates comprising colloidal materials (the plasma) and sand particles (skeleton grains) is enhanced by the enmeshment of microbial polysaccharides and polyuronide gums, fungal hyphae and fine plant rootlets.

The naturally formed aggregates or *peds* have *spheroidal*, *blocky*, *platy* or *prismatic* shapes, reflecting quite strongly the predominant pedogenic processes in the soil. The size, shape and arrangement of the peds determines the intervening void volume of cylindrical and spherical *pores* and planar cracks or *fissures*. Plasma concentrations in the pores and on ped faces resulting from the movement of materials in solution or suspension appear as a variety of coatings such as clay coatings (*cutans*), organic coatings (*organans*) and iron oxide coatings (*ferrans*).

The pore volume determines the size of the reservoir of water sustaining plant and animal life in the soil. Changes in water tension consequent upon changes in the *water-filled porosity* θ not only affect life processes but also have a marked bearing on structure formation itself. The pore volume not occupied by water defines the *air-filled porosity* ε, for which in a well aerated soil, the O_2 partial pressure is approximately the same as in the atmosphere, and that of CO_2 lies between 0.001 and 0.015 bars.

REFERENCES

Brewer R. (1964) *Fabric and Mineral Analysis of Soils*. Wiley, New York.
Kubiena W.L. (1938) *Micropedology*. Collegiate Press, Ames.
Soil Survey Manual (1951) *United States Department of Agriculture Handbook No. 18*.

FURTHER READING

Bullock P. & Murphy C.P. (1976) The microscopic examination of the structure of subsurface horizons of soils. *Outlook on Agriculture* 8, 348–354.
Edwards C.A. & Bremner J.M. (1967) Microaggregates in soils. *Journal of Soil Science* 18, 64–73.
Emerson W.W. (1959) The structure of soil crumbs. *Journal of Soil Science* 10, 235–244.
Harris R.F., Chesters G. & Allen O.N. (1966) Dynamics of soil aggregation. *Advances in Agronomy* 18, 107–169.
Hodgson J.M. (1974) (Ed.) Soil survey field handbook. *Soil Survey of England and Wales, Technical Monograph No. 5*, Harpenden.

Part 2
Processes in the
Soil Environment

'And earth is so surely the food of all plants that
with the proper share of the elements, which each
species requires, I do not find but that any
common earth will nourish any plant.'
Jethro Tull (1733) in
Horse-hoeing Husbandry,
republished by William Cobbett in 1829

Chapter 5
Soil Formation

5.1 THE SOIL-FORMING FACTORS

In 1941 Jenny, writing in his book *Factors of Soil Formation* presented an hypothesis which drew together many of the current ideas on soil formation, the inspiration for which owed much to the earlier studies of Dokuchaev and the Russian school, namely, that soil is formed as a result of the interaction of many variables, of which the most important are:

Parent Material
Climate
Organisms
Relief
Time

Jenny's approach to studying this interaction was firstly to establish the relationship between a soil property *s*, and any one of these variables, the others being held constant. The total change in *s* was then the sum of the separate changes in *s* induced by each of the soil-forming factors acting independently. Others have pointed out the shortcoming of this approach, in that parent material, climate etc. are not truly independent variables—for example, the operative climatic factor is not the regional climate but the microclimate existing in, and immediately above, the soil itself. This microclimate is modified by the rock strata, which affect the rate of leaching; by relief which affects runoff and drainage and by vegetation which in turn is very dependent upon the regional climate. Furthermore, climate, relief and so on rarely remain constant during the whole life of the soil.

Despite its original imperfections, Jenny's concept of a functional relationship of the kind:

$$s = f \text{ (parent material, climate, organisms, relief, time)} \tag{5.1}$$

is sound and lends itself to an analytical approach whereby, with the aid of a computer, variations in a soil property *s* can be quantified for two or more of the soil-forming factors changing simultaneously. The difficulty in practice is to assign meaningful values to the variables on the RHS of equation (5.1) for a single property *s*, remembering that the soil has many such properties for any one of which the functional relationship described by equation (5.1) may be unique. Nevertheless, it is instructive to discuss, individually, the factors—parent material, climate, organisms, relief and time—and their role in soil formation.

5.2 PARENT MATERIAL

Soil may form directly by the weathering of consolidated rock *in situ* (a residual soil), or it may develop on superficial deposits, which may have been transported by ice, water, wind or gravity. These deposits originate ultimately from the denudation and geologic erosion of consolidated rock.

Weathering

Generally, weathering proceeds by physical disruption of the rock structure which facilitates chemical changes within the constituent minerals. Forces of expansion and contraction induced by diurnal temperature variations cause rock shattering and exfoliation, as illustrated by surface features in a desert soil (Figure 5.1). This effect is most severe when water trapped in the rock repeatedly freezes and thaws, resulting in irresistable forces of expansion and contraction.

Water is the dominant agent in weathering, not only because it initiates solution and hydrolysis (section 9.2), but also because it sustains plant life on the rock surface. Lichens play a special part in weathering, since they are rich in chelating agents which trap the elements of the decomposing rock in organo-metallic complexes. In general, the growth, death and decay of plants and other organisms markedly enhances the solvent action of rainwater by the addition of carbon

Figure 5.1. Stones of a desert soil surface fractured due to extreme variations in temperature (J.J. Harris Teall in Holmes, 1978.

dioxide from respiration. The plant roots also aid in the physical disintegration of rock.

Water carrying suspended rock fragments has a scouring action on surfaces. The suspended material can vary in size from the finest 'rock flour' formed by the grinding action of a glacier, to the gravel, pebbles and boulders moved along and constantly abraded by fast-flowing streams. Particles carried by wind also have a 'sand-blasting' effect, and the combination of wind and water to form waves produces powerful destructive forces along shorelines.

Mineral stability

The rate of weathering depends on
(1) temperature;
(2) rate of water percolation;
(3) oxidation status of the weathering zone;
(4) surface area of rock exposed (largely determined by

physical processes of fracturing and exfoliation);
(5) types of mineral present.

Much attention has been paid to the relationship between the crystalline structure of a rock mineral and its susceptibility to weathering. One scheme for the order of stability, or *weathering sequence*, of the more common silicate minerals in *coarse* rock fragments (> 2 mm diameter) is shown in Figure 5.2. This sequence

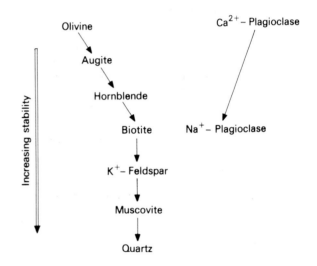

Figure 5.2. Sequence of silicate minerals in order of increasing stability (after Goldich, 1938).

is similar to that obtained by arranging the silicates in ascending order of the number of Si—O—Si bonds in the unit crystalline structure, which reflects the degree of polymerization of the silica (SiO_4^{4-}) tetrahedra. Generally speaking, the greater the degree of polymerization of the silica, the more resistant is the mineral to weathering.

Weathering continues in finely comminuted materials of soils and unconsolidated sedimentary deposits, following a sequence from the *least* to the *most stable* of minerals, the stages of which are identified by certain type minerals. Table 5.1 presents one such sequence, together with the broad characteristics of the associated soils.

Table 5.1. Stages in the weathering of soil minerals.

Stage	Type mineral	Soil characteristics
Early weathering stages		These minerals occur in the silt and clay fractions of young soils all over the world, and in soils of arid regions where lack of water inhibits chemical weathering and leaching.
1	Gypsum	
2	Calcite	
3	Hornblende	
4	Biotite	
5	Albite	
Intermediate weathering stages		Soils found mainly in the temperate regions of the world, frequently on parent materials of glacial or peri-glacial origin; generally fertile, with natural grass or forest vegetation.
6	Quartz	
7	Muscovite (also illite)	
8	Vermiculite and mixed layer minerals	
9	Montmorillonite	
Advanced weathering stages		The clay fractions of many highly weathered soils on old land surfaces of humid and hot intertropical regions are dominated by these minerals; often of low fertility.
10	Kaolinite	
11	Gibbsite	
12	Hematite (also goethite)	
13	Anatase	

(after Jackson *et al.*, 1948)

Rock types

Consolidated rocks are of igneous, sedimentary or metamorphic origin.

IGNEOUS ROCKS

These are formed by the solidification of molten magma in the Earth's crust, and are the ultimate source of all other rocks. They are broadly subdivided on mineral composition into *acidic* rocks—those relatively rich in quartz and the light-coloured Ca or K/Na silicates, and *basic* rocks—those low in quartz but high in the dark-coloured ferromagnesian minerals (hornblende, micas, pyroxene). The chemical composition of typical granite and basalt, rocks representative of these two groupings, is given in Table 5.2. The higher the ratio of Si to $(Ca + Mg + K)$, the more probable it is that some of the Si is not bound in the silicates, and may remain as residual quartz as the rock weathers. Soils formed from

Table 5.2. Chemical composition of typical granitic and basaltic rock.

	Granite	Basalt
	Per cent	
O	48.7	46.3
Si	33.1	24.7
Al	7.6	10.0
Fe	2.4	6.5
Mn	—	0.3
Mg	0.6	2.2
Ca	1.0	6.1
Na	2.1	3.0
K	4.4	0.8
H	0.1	0.1
Ti	0.2	0.6
P	—	0.1
Total	100.2	100.7

(after Mohr and van Baren, 1954)

granite are therefore usually higher in sand grains than those formed from basalt.

SEDIMENTARY ROCKS

These are composed of the weathering products of igneous and metamorphic rocks and are formed after deposition by wind and water. Cycles of geologic uplift, weathering and erosion, followed by deposition and gradual consolidation due to the weight of sediments above, have produced sequences of sedimentary beds which may be identified by their surface outcrops today. The *clastic* sedimentary rocks are composed of fragments of the more resistant minerals. The size of fragments decreases in order from the conglomerates to sandstones to siltstones to mudstones. Other sedimentary rocks are formed by precipitation or flocculation from solution—most commonly limestone and chalks (Figure 5.3) which vary in composition from pure calcium carbonate to mixtures of calcium and magnesium carbonate (dolomite), or carbonates and detrital material, usually flinty silica and clay minerals.

Igneous and sedimentary rocks that are subjected to intense heat and/or great pressures are transformed into *metamorphic* rocks. Changes in mineralogy and rock structure generally render the metamorphosed rock more resistant to weathering.

Figure 5.3. Electron micrograph of a sample of Middle Chalk (Berkshire, England) showing microstructure of calcified algae (courtesy of S.S. Foster).

Soils on transported material

SOIL LAYERS AND BURIED PROFILES

Most parts of the Earth's surface have undergone several cycles of submergence, uplift, erosion and denudation over the hundreds of millions of years of geological time. During the mobile and depositary phases there is much opportunity for the mixing of materials from different rock formations. Thus, the interpretation of soil genesis on such heterogeneous parent material is often complex. When two or more layers of transported material are superimposed they form a *lithological discontinuity*, and the resultant profile is often difficult to distinguish from that of a soil derived from consolidated rock *in situ*. In other cases, a soil may be buried by transported material during a period of prolonged erosion of an ancient landscape: the old soil is then preserved as a *fossil soil*, as illustrated by the buried mottled-zone of a laterite in Figure 5.4.

WATER

Transport by water produces alluvial, terrace and foot-slope deposits. During transport the rock material is

Figure 5.4. 'Mottled zone' of a fossil laterite preserved under more recent transported material in southern New South Wales.

sorted according to size and density and abraded, so that fluviatile deposits characteristically have smooth, round pebbles. The larger fragments are moved by rolling or by the process of *saltation* whereby the fragments are bounced along the stream bed by the force of water. The smallest particles are carried in suspension and salts move in solution. Colloidal material may not be deposited until the stream discharges into the sea, when flocculation occurs and estuarine deposits form. The modes of sediment transport in relation to the depth and velocity of the water are shown in Figure 5.5.

ICE

Ice was an important agent in the transport of rock materials during the 2 million years of the Pleistocene. During the colder *glacial* phases, the ice cap advanced from the North Pole and mountainous regions to cover a large part of the land surface in the Northern Hemisphere (the ice reached a line only a few kilometres north of the Thames Valley in the Midlands of Britain). The moving ice ground down rock surfaces and collected much debris on its surface by colluviation (see

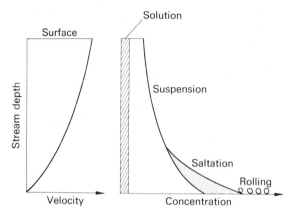

Figure 5.5. Sediment transport in relation to depth and velocity of water (after Mitchell, 1976).

below). During the warmer *interglacial* phases, the ice retreated leaving extensive deposits of *glacial drift*, including boulder clay and till, which is very heterogeneous in nature (Figure 5.6). Streams flowing out of the glaciers produced glacifluvial deposits of sands and gravels. Repeated freezing and thawing in periglacial regions resulted in frost-heaving of the surface and much swelling and compaction in subsurface layers. Such deposits may be distinguished from water-borne residues by the vertical orientation of many of the pebbles.

WIND

Wind moves rock fragments by processes of rolling, saltation and aerial dispersion. Material swept from dry periglacial regions has formed aeolian deposits called *loess* in central U.S.A., central Europe and China. These deposits, often cemented by the subsequent precipitation of calcium carbonate, are many metres thick in places and form the parent material of highly productive soils. Other characteristic wind deposits are the sand dunes of shorelines and desert areas.

GRAVITY

Gravity produces colluvial deposits, of which the scree formed by rock slides in mountainous regions is an obvious example. Less obvious, but widespread under periglacial conditions, is *solifluction* whereby wet thawed soil slips over the frozen ground below, even on

Figure 5.6. Soil formation on glacifluvial sands and gravels of the Lleyn Peninsular, Wales.

gentle hillslopes; and *creep* whereby the surface mantle of soil moves by slow colluviation away from the original rock. The junction between creep layer and weathering rock is frequently delineated by a stone line.

5.3 CLIMATE

The key components of climate in soil formation are moisture and temperature.

Moisture

The effectiveness of moisture depends on several factors:
(1) the form and intensity of the precipitation;
(2) its seasonal variability;
(3) the evaporation rate (from vegetation and soil, see section 6.6);
(4) land slope;
(5) permeability of the parent material.

Various functions of meteorological variables have been proposed to measure *moisture effectiveness*. One such function is Thornthwaite's (1931) *P–E* index, which is the summation over one year of the monthly ratios:

$$\frac{\text{Precipitation } P}{\text{Evaporation } E} \times 10 \qquad (5.2)$$

(The multiplier of 10 was introduced to eliminate fractions.) Based on the *P–E* index, five major categories of climate and associated vegetation types are recognized, namely:

P–E index	Category	Vegetation
128 and above	wet	rainforest
64–127	humid	forest
32–63	subhumid	grassland
16–31	semiarid	steppe
< 16	arid	desert

When moisture effectiveness is high, as in wet or humid climates, there is a net downward movement of water in the soil for most of the year, which usually results in greater *leaching* of soluble materials, sometimes out of the solum, and the *translocation* of clay particles from upper to lower horizons. More will be said about these processes, and profile differentiation, in Chapter 9.

Temperature

Temperature varies with latitude and altitude, and the extent of absorption and reflection of solar radiation by the atmosphere. Temperature affects the rate of mineral weathering and synthesis, and the biological processes of growth and decomposition. Reaction rates are roughly doubled for each 10°C rise in temperature, although enzyme-catalysed reactions are sensitive to high temperatures and usually attain a maximum between 30 and 35°C.

Categories based on the thermal efficiency of the climate may be delineated by Thornthwaite's *T–E* index, which is the sum for one year of the monthly ratios:

$$\frac{\text{mean monthly temperature (°F)} - 32}{4} \qquad (5.3)$$

This empirical expression was chosen so that the cold margin of the Tundra had a *T–E* index of zero and the cooler margin of the rainforest an index of 128. The *T–E* indices and categories are:

T–E index	Category
128 and above	tropical
64–127	mesothermal
32–63	microthermal
16–31	taiga
1–15	tundra
0	frost

Zonal soils

From Dokuchaev (*c.* 1870) on, many pedologists in Europe and North America have regarded climate as pre-eminent in soil formation. The obvious relationship between climatic zones, with their associated vegetation, and the broad belts of similar soils which stretched roughly east–west across Russia inspired the zonal concept of soils.

Zonal or normal soils are those in which the climatic factor, acting on the soil for a sufficient length of time, is so strong as to override the influence of any other factor.

Intrazonal soils, although showing well developed soil features, are those in which some local anomaly of relief, parent material or vegetation is sufficiently strong to modify the influence of the regional climate.

Azonal or immature soils have poorly differentiated

profiles, either because of their youth or because some factor of the parent material or environment has arrested their development. The zonal concept gave rise to systems of classifying soils reflecting the known or presumed genesis of like soils—the *Great Soil Groups* as listed in Table 5.3. Such a scheme was adopted, with minor modifications, over large continental areas,

notably the U.S.A. and Australia, and although the classification is now superseded, the use of the Great Soil Group names remains common practice.

Predictably, the concept of soil zonality is not very helpful when applied to soils of the subtropics and Tropics. There, where land surfaces are generally much older than in Europe and North America, and have

Table 5.3. Higher soil categories according to the Zonal Concept.

Order	Suborder	Great soil groups
Zonal soils	1. Soils of the cold zone	Tundra soils
	2. Light-coloured soils of arid regions	Desert soils Red desert soils Sierozems Brown soils Reddish-brown soils
	3. Dark-coloured soils of semiarid, subhumid, and humid grasslands	Chestnut soils Reddish chestnut soils Chernozems Prairie soils Reddish prairie soils
	4. Soils of the forest-grassland transition	Degraded chernozems Noncalcic brown or Shantung brown soils
	5. Light-coloured podzolised soils of the forest regions	Podzols Gray wooded Brown podzolic soils Gray-brown podzolic soils Red-yellow podzolic soils
	6. Lateritic soils of forested mesothermal and tropical regions	Reddish-brown lateritic soils Yellowish-brown lateritic soils Laterites
Intrazonal soils	1. Halomorphic (saline and alkali) soils of imperfectly drained arid regions and littoral deposits	Solonchak or saline soils Solonetz Soloth (solod)
	2. Hydromorphic soils of marshes, swamps, seep areas and flats	Humic gley soils Alpine meadow soils Bog soils Half-bog soils Low-humic gley soils Planosols Groundwater podzols Groundwater laterites
	3. Calcimorphic soils	Brown forest soils (Braunerde) Rendzinas
Azonal soils		Lithosols Regosols (includes Dry sands) Alluvial soils

(after the USDA Soil Survey Staff, 1960)

consequently undergone many cycles of erosion and deposition associated with climatic change, the age of the soil (section 5.6) and its topographical relation to other soils in the landscape (section 5.5) are factors of major importance.

5.4 ORGANISMS

The soil, and the organisms living on it and in it, comprise an *ecosystem*. The active components of the soil ecosystem are the plants, animals, microorganisms and man.

Vegetation
The primary succession of plants which colonize a weathering rock surface (section 1.1) culminates in the development of a *climax community*, the species composition of which depends on the climate and parent material, but which, in turn, has a profound influence on the soil that is formed. For example, in the mid-West and Great Plains of the United States, deciduous forest seems to accelerate soil formation compared to prairie (grassland) on the same parent material and under similar climatic conditions. In parts of England such as on the North York moors or the Bagshot Sands, the profiles preserved under the oldest barrows (burial mounds) of the Bronze Age people are typical of *brown forest soils*, formed under the deciduous forests of some 4000 years ago, in contrast to the *podzols* on the surfaces of the barrows which have formed under the usurping heathland vegetation. Differences in the chemical composition of leaf leachates can partly account for this divergent pattern of soil formation (section 9.2).

Fauna
The species of invertebrate animals in the soil and the type of vegetation are closely related. The acid litter of pines, spruce and larch creates an environment unfavourable to earthworms, so that litter accumulates on the soil surface where it is only slowly comminuted by mites and *Collembola* and decomposed by fungi. The litter of elm and ash, and to a lesser extent oak and beech, is more readily ingested by earthworms and becomes incorporated into the soil as faecal material.

Earthworms are the most important of the soil-forming fauna in temperate regions, being supported to a variable extent by the small arthropods and the larger burrowing animals (rabbits and moles). Earthworms are also important in tropical soils, under good rainfall, but in general the activities of termites, ants and coprophagous (dung-eating) beetles are of greater significance, particularly in the subhumid to semiarid savanna of Africa and Asia.

Earthworms of the casting type build up a stone-free layer at the soil surface, as well as intimately mixing the litter with fine mineral particles they have ingested. The surface area of the organic matter that is accessible to microbial attack is then much greater. Termites are also important soil-formers in that they carry fine particles, water and dissolved salts from depths of 1 metre or more to their termitaria on the surface. On abandonment, the termitaria are reduced by weathering and the trampling of animals to form a thin mantle on the soil.

Man
Man influences soil formation through his manipulation of the natural vegetation, and its associated animals, his agricultural practices and his urban and industrial development. Indigenous populations of wild animals are regulated by natural controls; not so domestic stock which can, if present in excessive numbers, so denude the vegetation that the soil is exposed to severe erosion risk. Compaction produced by the traffic of domestic and wild animals decreases the rate of water infiltration into the soil, thereby increasing problems of runoff and erosion. The subject of man and his management of soil resources is discussed in Part III.

5.5 RELIEF

Relief has an important influence on the local climate, vegetation and drainage of a landscape. Major topographical features are easily recognized—mountains and valleys; escarpments, ridges and gorges; hills, plateaux and floodplains. Appreciable changes in elevation affect the amount and form of the precipitation, the intensity of storm events, as well as the type of vegetation. More subtle changes in local climate and vegetation are associated with the *slope* and *aspect* of valley sides and escarpments.

Slope

Angle of slope and vegetative cover affect moisture effectiveness by governing the proportion that surface runoff of water bears to infiltration. Soils with impermeable subsoils, and those developing on slopes, may also show appreciable lateral sub-surface flow. Thus, at the top of a slope, the soils tend to be freely drained with the water table at considerable depth, whereas at the valley bottom the soils are poorly drained, with the watertable near or at the soil surface (Figure 5.7). The succession of soils forming under different drainage conditions but on relatively uniform parent material comprises a *hydrological sequence*.

In addition to modified drainage, there is movement of soil on slopes; a mid-slope site is continually receiving solids and solutes from sites immediately up-slope by wash and creep, and continually losing material to sites below. Here the *form* of the slope is important—whether smooth or uneven, convex or concave, or broken by old river terraces. Minor variations in relief such as those caused by frost-heaving in very cold regions, or by faunal activity (rabbits, moles and termites) promote the slow creep of soils down even slight slopes.

Small variations in elevation can also be important in flat lands, especially if the groundwater is saline and rises close to the surface. A curious pattern of micro-relief of this kind, called *gilgai**, has been observed in the chernozem-like soils of semiarid regions of Australia (Figure 5.8). The soils of the mound ('puff') and hollow ('shelf') are quite different, with calcium carbonate and gypsum occurring much closer to the surface in the yellow-brown crumbly clay of the puff than in the grey, compact cracking clay of the shelf. Similar features have been reported in the black clay soils of Texas, the Orange Free State and in the Black Cotton soils of the Deccan.

The catena

The effect of relief on soil formation has been studied less than the destruction of soil on slopes by erosion. This neglect reflects the assumption of many early pedologists that a soil forming on other than a gently undulating surface did not develop a normal (zonal)

* from the Aboriginal word for a small waterhole

Figure 5.7. Section of slope and valley bottom showing a hydrological soil sequence.

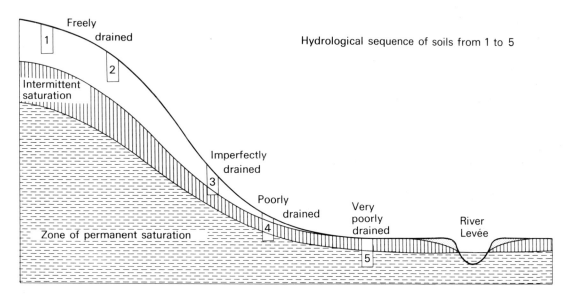

Hydrological sequence of soils from 1 to 5

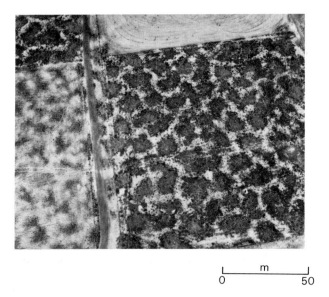

```
        L_____m_____J
        0              50
```

Figure 5.8. Aerial view of gilgai micro-relief (CSIRO photograph courtesy of G.D. Hubble).

profile, and was therefore classed as intrazonal or azonal. However, the importance of relief was highlighted by Milne (1935) in central East Africa, who recognized a recurring sequence of soils forming on slopes on similar parent materials. He named the slope sequence a *catena* (from the Latin for chain). Uniformity of parent material is of secondary importance in the very old landscapes of the Tropics, but not in temperate regions subjected to recent glaciation with consequent 'rejuvenation' of parent materials.

From its original conception as a mapping unit, the meaning of 'catena' has been broadened by usage to include many denudational and hydrological sequences of soils on slopes, of which Figure 5.7 is one example; others appear in Chapter 9. In this sense, the term is synonymous with Jenny's *toposequence*, which defines the suite formed when the influence of relief is dominant over that of the other soil-forming factors.

5.6 TIME

Time acts indirectly on soil formation by determining the duration for which a given set of factors operates.

It is well known that the world climate has changed over geological time: the most recent, large changes were associated with the alternating glacial and inter-glacial periods of the Pleistocene. Major climatic changes were accompanied by the raising and lowering of sea level, erosion and deposition, and induced isostatic processes in the Earth's crust, all of which produced radical changes in the distribution of parent material and vegetation, and in the shape of the landscape. On a much shorter timescale, covering a few thousand years, changes can occur in the biotic factor of soil formation, as witnessed by the primary succession of species on a weathering rock surface; and also in relief, as exemplified by changes in slope form and the distribution of groundwater. In the soil, the stage reached in the weathering sequence by a rock fragment is dependent on the intensity of weathering and the time for which it has occurred.

Profile features

One criterion of a soil's age is *profile morphology*, in particular, the multiplicity of horizons, their degree of differentiation, and their depth. The use of profile criteria is feasible for a soil derived from uniform parent material. Frequently, however, the past history of climatic, geologic and geomorphological change is so complex that an accurate assessment of a soil's age is impossible. Where organic matter occurs at depth, the method of carbon dating using the natural radio-activity of carbon can be applied. The relation between soil development and time can also be determined in special cases where the time elapsed since the exposure of parent material is known, and the subsequent influence of the other soil-forming factors has been reasonably steady. Such studies as have been made suggest that the rate of soil development is extremely variable, ranging from very rapid (less than 100 years) on volcanic ash in the Tropics to very slow (1 cm in 5000 years) on chalk weathering in a cool temperate environment.

Soil maturity

The extrapolation of contemporary measurements, such as those mentioned above, to assess soil age must be tempered by the realization that soil formation

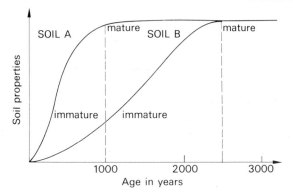

Soils A and B formed on different parent materials

Figure 5.9. Soil formation with time under a cool temperate climate (after Jenny, 1941).

follows a 'law of diminishing returns'. That is to say, the soil attains a steady-state equilibrium with the environment, in which the rate of soil formation is equal to its rate of destruction (Figure 5.9). This condition corresponds to a *mature soil*. Mature soils

Figure 5.10. Soil formation with time on volcanic ash under a wet tropical climate (after Mohr and van Baren, 1954).

are, by implication, stable soils and support stable ecosystems. One of the problems which has confronted agriculturalists for centuries is how to preserve a soil whilst utilizing it for crop production, an activity which invariably changes the position of the soil-environment equilibrium.

However, situations also arise under natural conditions where the rate of destructive processes exceeds the rate of accretion and retention of materials from rock weathering, plant growth and aerial deposition. The trend in soil formation with time on recent volcanic ash in the Tropics (Figure 5.10) shows that soil depth and profile differentiation attain maxima in the virile stage, only to decline in the senile and final stages. The crucial difference between this, and the example of Figure 5.9, lies in the temperature and the rate of water percolation through the soil. The gains and losses in the soil formed by intense weathering and leaching of the highly permeable volcanic ash do not achieve a balance, and the final equilibrium state is indissociable from the virtual destruction of the soil itself.

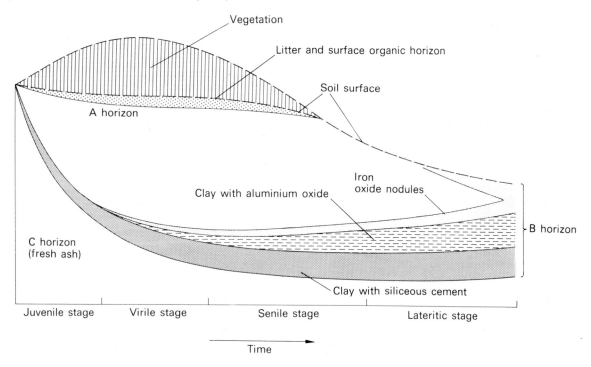

5.7 SUMMARY

Soil formation is an integrated process, the direction and pace of which depend on the influence of *parent material*, *climate*, *organisms* and *relief*, and the length of *time* these factors have interacted. Water is the major agent of rock weathering, its solvent action being greatly enhanced by the respiration of plants and microorganisms. The residues of weathering are redistributed by the action of wind, water, ice and gravity.

For those land surfaces left covered by glacial drift, outwash material and periglacial wind-blown deposits at the end of the Pleistocene, some 11,000 years ago, the influence of parent material and relief is often subordinate to that of recent climate, and the associated climax vegetation—the soils formed are essentially *monogenetic*. The important components of the climate are *temperature* and *moisture effectiveness*, the latter depending primarily on the regional balance between precipitation and evaporation.

In the warmer intertropical regions where the greater part of the land surface is much older than in temperate regions, the pattern and characteristics of past cycles of soil formation are often preserved in deep mantles of weathered material—the profiles are *polygenetic* in origin. In such regions, parent material is important where erosion is sufficiently active to remove the products of weathering; otherwise relief and time of exposure to weathering are the major factors in soil formation.

REFERENCES

JENNY H. (1941) *Factors of Soil Formation*. McGraw-Hill, New York.

MILNE G. (1935) Some suggested units of classification and mapping, particularly for East African soils. *Soil Research* **4**, 183–198.

THORNTHWAITE C.W. (1931) The climates of North America according to a new classification. *Geographical Review* **21**, 633–656.

FURTHER READING

BIRKELAND P.W. (1974) *Pedology, Weathering and Geomorphological Research*. Oxford University Press, New York.

CLARKE G.R. (1971) *The Study of the Soil in the Field*. Clarendon Press, Oxford.

HOLMES A. (1978) *Principles of Physical Geology*. 3rd Ed. Nelson, London.

MOHR E.C.J. & VAN BAREN F.A. (1954) *Tropical Soils*. N.V. Uitgeveriz, The Hague.

THOMAS M.J. (1974) *Tropical Geomorphology*. Macmillan, London.

Chapter 6
Soil Water and the
Hydrologic Cycle

6.1 THE FREE ENERGY OF SOIL WATER

Total water potential

A small quantity of water added to a dry soil is distributed so that the force of attraction between soil and water is as large as possible (achieving a maximum reduction in free energy). Additional water is held by progressively weaker forces so that eventually, when the soil is saturated, there is little difference between the free energy of the soil water and water in the standard state, which is taken as pure, free water at the same temperature, pressure and height above sea level as the water in the soil. The difference in free energy may be calculated, per mole of water, from the equation:

$$\mu_w \text{ (soil)} - \mu_w^\circ \text{ (standard state)} = RT \ln e/e^\circ \qquad (6.1)$$

where μ_w = chemical potential (free energy/mole) of water;
e/e° = ratio of vapour pressure of water in the soil to vapour pressure of water in the standard state;
R = gas constant;
and T = absolute temperature (°K). (The reference state for pressure is taken as atmospheric pressure.)

This difference in chemical potential defines the *total water potential* ψ. The upper value of ψ is obtained at $e/e^\circ = 1$, when $\psi = 0$. The units of ψ are energy per mole (J mol^{-1}), but to be compatible with earlier *pressure* measurements, ψ is usually expressed as energy per unit volume (J m^{-3}), since J m^{-3} is equivalent to Nm^{-2}.

The hydrostatic pressure p due to a column of water h metres high is given by

$$p = h \rho g \qquad (6.2)$$

where ρ = density of water (kg m^{-3}) and
g = acceleration due to gravity (m^{-2})

Thus, a pressure of 1 *bar* (10^5 N m^{-2}) is subtended at the base of a water column of height

$$\frac{p}{\rho g} = \frac{10^5}{1000 \times 9.8} = 10.2 \text{ m}$$

Hydrologists make use of this relationship and usually express pressure as the height of the equivalent water column, referred to as *hydraulic head* or simply *head*.

Soil water may be held under a positive pressure p, or negative pressure (tension) due to the application of a suction (Figure 6.1). In dry soils, the suction of water,

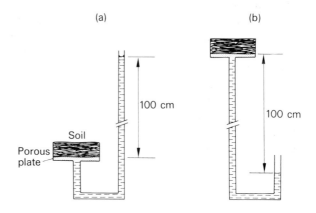

Figure 6.1. (a) Soil water under a hydrostatic pressure of 0.1 bar.
(b) Soil water under a negative hydrostatic pressure of 0.1 bar.

when expressed in head units in the centimetre-gram-second system, was often a large and unwieldy number. It was therefore more convenient to use $\log_{10} h$ in cm, which defines the *pF value* of the soil water. Comparative values of water potential, head and *pF* for a soil at various states of dryness are given in Table 6.1.

Components of soil water potential

The free energy of water in the soil is reduced in the following ways:
(a) Water is strongly adsorbed by hydrogen-bonding at the hydroxyl surfaces of kaolinites and silica minerals (the $\equiv$Si–OH grouping). The water molecules are highly oriented in the first adsorbed layer, the structure

Table 6.1. Water potential, hydraulic head and pF of a soil of varying wetness.

Soil moisture condition	Water potential, ψ	Head,	pF	Equivalent radius of largest pores that would just hold water
	(bars)	(m)		(μm)
Soil saturated or very nearly so	-0.001^*	0.0102^{**}	0	1500†
Moisture held after free drainage (field capacity)	-0.05	0.51	1.7	30
Approximately the moisture content at which plants wilt	-15	153	4.2	0.1
Soil at equilibrium with a relative vapour pressure of 0.85 (approaching air-dryness)	-220	2244	5.4	0.007

* calculated from equation 6.1
** calculated from equation 6.2
† calculated from equation 6.4

becoming less orderly and more like that of free water as the layers multiply. At the charged planar surfaces of the 2:1 type clay minerals, however, the water is strongly attracted to the exchangeable cations and only weakly bonded, if at all, to the oxygen surfaces. The electric field of the cation orients the polar water molecules around the ion to form a hydration shell (Figure 6.2), which is an example of an ion-dipole

Figure 6.2. A hydrated exchangeable cation.

interaction (section 7.5). The energy of hydration per mole of cation depends upon its ionic radius and

charge, the energies for the common exchangeable cations falling in the order:

$$Al^{3+} > Mg^{2+} > Ca^{2+} > Na^+ > K^+ \simeq NH_4^+$$

Of course, the greater the hydration energy of the cation, the larger is the reduction in free energy of the water molecules in its hydration shell.

Thus, the small amount of water present in very dry soils (at relative vapour pressures <0.5) is mainly associated with the exchangeable cations.

(b) As water layers build up, the water begins to fill the finest pores and interstices between mineral particles, forming curved air–water menisci (Figure 6.3a).

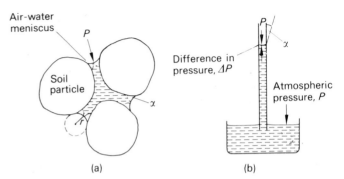

Figure 6.3. (a) Water held under tension between soil particles. **(b)** Rise of water in a capillary tube.

Equilibrium is attained when the surface free energy associated with the air–water–solid interface is at a minimum. The work done on the water is balanced by the increased potential energy of the water column, as for example in the rise of water in a capillary tube (Figure 6.3b). It follows from the latter figure that

$$\pi r^2 \rho g h = 2 \pi r \gamma \cos\alpha \qquad (6.3)$$

where π, r, ρ, g, h and α are as already defined, or indicated in Figure 6.3, and γ is the surface tension of water. Thus, the difference in pressure Δp across the meniscus is given by

$$\Delta p = \rho g h = \frac{2\gamma}{r} \qquad (6.4)$$

when the angle of wetting $\alpha = 0$.

Relative to atmospheric pressure, the pressure difference across the air–water interface in the soil pore is numerically equal to the *matric potential* (ψ_m) of the pore water. Water held in pores of diameter 1 μm, for example, has a matric potential of -3 bars.

(c) Solutes (ions and organic molecules) also reduce the free energy of soil water, the extent of the reduction being measured by the *osmotic potential* ψ_s. Osmotic potential is numerically equal to the osmotic pressure of the soil solution, which is defined as the hydrostatic pressure necessary to just stop the inflow of water when the solution is separated from pure water by a semipermeable membrane.

ψ_m and ψ_s are *component potentials* of the soil water potential ψ. In soil they are always less than 0 so that ψ is also negative, unless the effect of ψ_s and ψ_m is outweighed by contributions from *either* water held at a height above that chosen for the standard state, giving rise to a positive *gravitational potential* (ψ_g), *or* water under a pressure greater than atmospheric, giving rise to a positive *pressure potential* (ψ_p).

The soil water potential ψ is therefore given by the sum

$$\psi = \psi_m + \psi_s + \psi_p + \psi_g \qquad (6.5)$$

The retention and movement of water in the soil is discussed in terms of these potentials in the succeeding sections; for soil-plant relations in general, however, the important components of ψ are the matric, osmotic and gravitational potentials.

6.2 THE HYDROLOGIC CYCLE

A global picture
Water vapour enters the atmosphere by evaporation from soil, water and plants; it precipitates as rain, hail, snow or dew. Precipitation P equals evaporation E over the entire Earth's surface, but for land surfaces only, the average annual P of 0.71 m exceeds the average annual E by 0.24 m, the difference amounting to river discharge. If that part of the land surface which contributes little to evaporation—the deserts, ice caps and Tundra—is excluded from the calculation, the average rate of evaporation from the remainder is approximately 1 m per annum, which is not far short

of the 1.2 m annual rate of evaporation from the oceans.

Evaporation depends on an input of energy, the energy available being much greater at the Equator than at high latitudes. The capacity of the air to hold water also increases with temperature, so that the amount of precipitable water is greatest at the Equator, least at the Poles, and more in summer than in winter (Figure 6.4). It is less easy to estimate accurately the

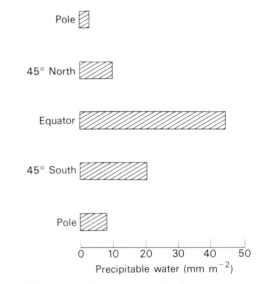

Figure 6.4. Amount of water vapour in the atmosphere by latitude (after Penman, 1970).

larger quantities of surface water, soil water and groundwater which sustain the terrestrial plant and animal life: the average equivalent depth of soil water is *c.* 0.25 m, that of lake and river water *c.* 1 m and groundwater between 15 and 45 m.

Water balance on a local scale
As illustrated in Figure 6.5, of the gross precipitation on land, nearly all is intercepted by the vegetation with the remainder falling directly onto the soil. Water in excess of that required to wet the leaves and branches— the *canopy storage capacity* (usually 0.5–1 mm), drips from the canopy or runs down stems to the soil. Canopy drip, stem flow and direct rainfall constitute the *net precipitation* or *net rainfall*. Water retained by the canopy is lost by evaporation and is referred to as

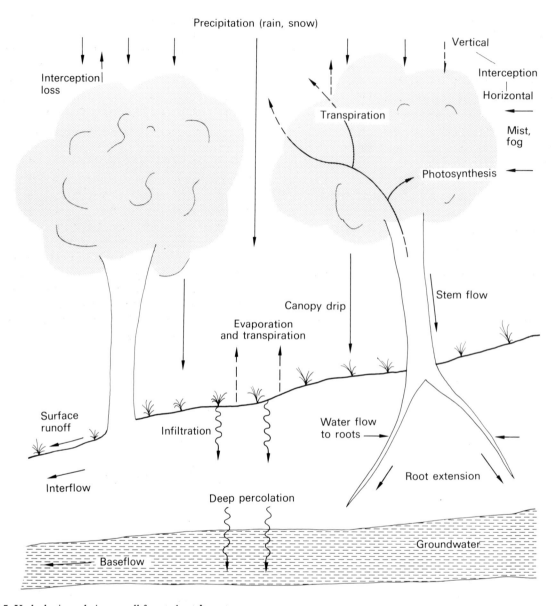

Figure 6.5. Hydrologic cycle in a small forested catchment.

interception loss: it may be as much as one-quarter of the gross rainfall for broadleaf trees in summer.

Of the rain which reaches the soil, the difference between that soaking in and that running off comprises *surface runoff* or *overland flow*. The movement of water into the soil from above is called *infiltration*. When the soil is thoroughly wet down to the watertable, a steady rate of flow or *percolation* to the groundwater is maintained. Subsurface lateral flow or *interflow* also occurs through soils on slopes, or when vertical flow through the subsoil is impeded. Interflow eventually contributes to streamflow from the catchment, as does the lateral flow of groundwater which is called *baseflow*. Surface runoff makes a much less predictable contribution to streamflow and usually is accompanied by soil erosion.

Slow upward movement of water occurs in response to evaporation from the soil surface. With abundant plant growth, however, some water moves into the water-depleted zones around the roots and much is intercepted by the roots growing through the soil, to be transported to the aerial parts and lost by evaporation through the leaves—the process of *transpiration*.

The water balance equation
The interaction of all these processes within a self-contained catchment is expressed by the equation:

$P = E +$ Streamflow $\pm$ Deep percolation $\pm$ Change in soil storage, ΔS (6.6)

For catchments with impermeable substrata, when periods of one year or more are considered, the changes in soil storage ($\pm \Delta S$) tend to cancel out so that the difference between P and E is primarily due to streamflow. Over shorter periods, values of ΔS are nevertheless important for plant growth, and if ΔS is negative irrigation may be necessary to supplement rainfall.

A knowledge of streamflow and deep percolation in a catchment is also important, firstly because streamflow provides water that can be conserved for human and animal use; secondly because water in deep aquifers (permeable rocks that hold groundwater) is a valuable reserve from which to supplement surface supplies when E is consistently greater than P. Deep percolation is largely beyond man's control, as is precipitation, despite attempts made in the U.S.A. and Australia to induce rainfall by 'cloud seeding' with dry ice or silver iodide; there are, however, opportunities for manipulating the other variables in equation (6.6) which depend to a large extent on soil properties and land management in the catchment. It is therefore relevant to discuss infiltration, soil water movement and evaporation in some detail.

Figure 6.6. Hydraulic head values for water in soil under a positive pressure p or suction s (after Youngs and Thomassen, 1975).

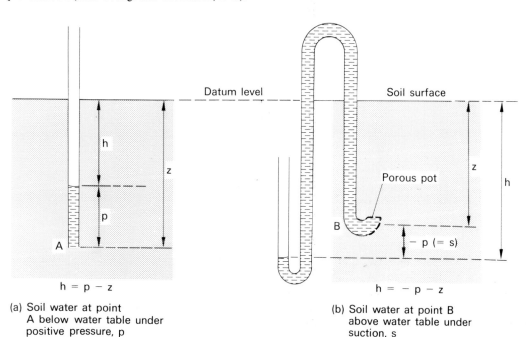

(a) Soil water at point A below water table under positive pressure, p

(b) Soil water at point B above water table under suction, s

6.3 INFILTRATION

Darcy's law

For water movement through porous materials, provided the velocity is low enough that flow is laminar and not turbulent, the rate of flow is directly proportional to the driving force and inversely proportional to the resistance, as expressed by Darcy's equation:

$$v = -k \text{ grad } \psi \qquad (6.7)$$

where v is the volume per unit time crossing an area A perpendicular to the flow; $grad\ \psi$ denotes the change in water potential per unit distance in the direction of flow (which may be horizontal or vertical); and k is a constant called the *hydraulic conductivity* which is defined as the reciprocal of the flow resistance. (The negative sign is a convention indicating that flow occurs in the opposite direction to the potential gradient.)

The value of k depends on the amount of water in the pores and its viscosity, as well as on the pore geometry and surface roughness. The ideal conditions for applying Darcy's law are that both $grad\ \psi$ and the resistance to flow are constant. Soils rarely satisfy these conditions because they are in a dynamic cycle of wetting and drying so that the parameters in equation 6.7 are continually changing in space and time (section 6.4).

For all except saline soils, ψ_m and ψ_g are the main component potentials governing water movement. Written in head units, the total water potential becomes the sum of the pressure head p, where $-p = $ suction s for soil water under tension, and the gravitational head z, measured positively upwards from an arbitrary datum level, usually the soil surface, as illustrated in Figure 6.6. Thus,

$$h = p + z \qquad (6.8)$$

and equation 6.7 becomes

$$v = -k \text{ grad } h \qquad (6.9)$$

which for flow downwards through soil above the water table may be written

$$v = -k\left(-\frac{dp}{dz} - \frac{dz}{dz}\right) = -k\left(\frac{ds}{dz} - 1\right) \qquad (6.10)$$

Stages of infiltration

Rain falling on the soil surface is drawn into the pores under the influence of both a suction and gravitational head gradient, and if the rainfall intensity is less than the initial *infiltration rate* (*I.R.*) all the water is absorbed. In the early stages of wetting a dry soil, the suction gradient is predominant. However, as the antecedent soil water content usually increases with depth, the suction gradient decreases as the water penetrates deeper into the soil. This effect, together with structural changes in the surface following raindrop impact which reduce the available porosity, cause the *I.R. to fall below* the rainfall intensity within a few minutes to perhaps hours. At this stage ponding of water in surface depressions occurs, leading to surface runoff. When the whole soil profile has been rewet, water flows steadily downwards in response to the gravitational head gradient, and a constant *I.R.*—the *infiltration capacity* or *infiltrability*—is observed. This sequence of events for rainfall of moderate intensity is illustrated in Figure 6.7.

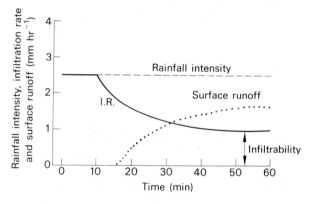

Figure 6.7. Infiltration and surface runoff during a storm.

6.4 REDISTRIBUTION OF SOIL WATER

Saturated flow

THE WETTING FRONT

Water moving downwards into a homogeneous dry soil travels at a constant velocity to form a *transmission zone* behind a narrow wetting zone and well defined *wetting front.* The plot of water content against depth for the soil profile, illustrated in Figure 6.8, shows a sharp

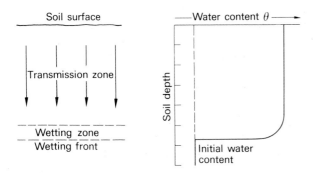

Figure 6.8. Vertical infiltration into a homogeneous column of dry soil.

change in water content at the wetting front. This occurs because water at the boundary takes up a preferred position of minimum potential in the narrowest pores, for which the hydraulic conductivity is very low, and does not move at an appreciable rate until the macropores begin to fill. However, owing to parent material heterogeneity and the effects of soil fauna and plant roots, field soils are rarely as regularly structured as Figure 6.8 would imply, so that downward movement of water is more erratic, as illustrated by the wetting front of the soil profile in Figure 6.9.

Figure 6.9. Vertical infiltration into a soil in the field.

As long as water is applied to the soil surface, the transmission zone is near-saturated (allowing for some trapped air pockets) and the water flow is described by Darcy's law. Provided that the water has penetrated

fairly deeply (>0.5 m), the suction gradient ds/dz is small and the term $-k\,ds/dz$ in equation 6.10 is negligible, so that $v=k$, the *saturated permeability* of the soil (which will also be the infiltrability). The range of values for saturated permeability is 1–500 mm day^{-1}.

FIELD CAPACITY

When infiltration ceases, redistribution of water occurs at the expense of the initially saturated zone of soil. If the soil is completely wet to the watertable, or an artificial outlet such as a drain, drainage ceases when the suction gradient acting upwards balances the gravitational head gradient acting downwards. Even when the soil is not completely wet, depthwise, the drainage rate often becomes very small one to two days after rain or irrigation. This is possible because of the sharp fall in the hydraulic conductivity of the soil when most of the macropores have drained of water. The water content then defines the *field capacity* (*FC*), which for well structured British soils is in equilibrium with a suction of 50 mbars (section 4.5). As we shall see in section 6.5, the concept of field capacity is useful for setting an upper limit to the amount of 'available' water in the soil; but it is imprecise—firstly because the soil may remain above the *FC* due to frequent showers of rain for several days, and secondly because soils with a very high proportion of micropores continue to drain slowly for several days after rain.

Unsaturated flow

STEADY-STATE CONDITIONS

As the large pores empty, the hydraulic conductivity of the soil falls and water flow is said to be unsaturated. However, if the soil water content stabilizes at a lower value—as occurs when the rate of inflow into a given volume matches the outflow rate, described as a *steady-state* condition—Darcy's equation is still applicable in the form:

$$v = -k\left(\frac{ds}{dz} - 1\right) \qquad (6.11)$$

where k is now the *unsaturated permeability* of the soil. An example of such flow occurs when rainfall of constant intensity, less than the saturated permeability of the soil, percolates to a watertable at a fixed depth below the surface.

For drainage to a very deep water table ds/dz is much smaller than 1, and the water content throughout the profile adjusts so that $v=k$ (compare with saturated flow through a deep transmission zone during infiltration). Unsaturated flow of this kind occurs in deep permeable soils below the roots of the vegetation. Alternatively, flow is sometimes impeded by a layer of less permeable soil or parent material deep in the profile, resulting in a *perched water table* (Figure 6.10).

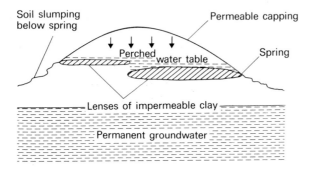

Figure 6.10. A landscape profile showing perched water tables and spring lines.

The increase in hydraulic head immediately above the impeding layer induces lateral flow which may eventually appear as a line of springs at another position in the landscape.

NON-STEADY STATE CONDITIONS

Soil water is rarely in steady-state equilibrium in the root zone. Intermittent rainfall and surface evaporation combined with the variable plant uptake of water causes the water content θ to fluctuate from point to point with time. The flow can then only be described by combining Darcy's equation with the continuity equation, which relates the rate of change of θ in a small soil volume to the change in flow rate into and out of that volume, allowing for any water removed by evaporation and transpiration (q) (Figure 6.11). Accordingly,

$$\frac{d\theta}{dt} = -\frac{dv}{dz} - q.$$

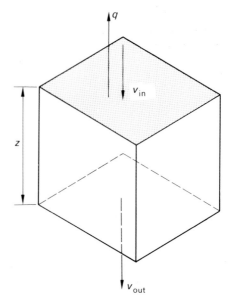

Figure 6.11. Water flow through a soil volume under non-steady state conditions.

Substituting for v from equation 6.11 we have

$$\frac{d\theta}{dt} = \frac{d}{dz}\left\{ k\frac{ds}{dz} - k \right\} - q \qquad (6.12)$$

Suction s can be measured using tensiometers, at values <0.85 bars, and by gypsum blocks or soil psychrometers at higher suctions; θ can be determined *in situ* with a neutron probe or by sampling the soil, oven-drying and weighing. If values for q can be estimated (section 6.6), equation 6.12 can be solved for k. Further details are given in advanced texts by Childs (1969) and Hillel (1971).

6.5 THE MOISTURE CHARACTERISTIC CURVE

Pore size distribution

If a saturated soil core is placed on a porous plate attached to a hanging water column and allowed to attain equilibrium (Figure 6.1b), the soil water content at the applied suction can be determined. For suctions >0.85 bars, the soil is usually brought to equilibrium with an air pressure which is then equal but opposite to the water suction in the soil pores. A graph of θ against soil suction is constructed, which is called the *moisture characteristic* or *moisture retentivity curve*, illustrated by curve A in Figure 6.12. The rate of

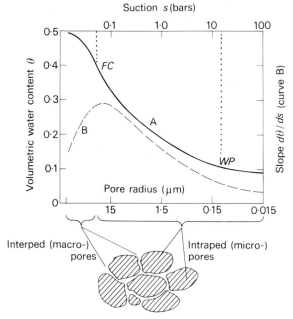

Figure 6.12. Graphs of the moisture characteristic (A) and specific water capacity (B) of a clay soil.

change of θ with s ($d\theta/ds$) is the *specific* or *differential water capacity*.

Since s is inversely proportional to the radius of the largest pores holding water at that suction (equation 6.4), the volume of pores having radii between r and $r - \delta r$, where δr is a very small decrement in r corresponding to an increase in suction δs, can be determined from the slope of the moisture characteristic curve, provided that shrinkage of the soil on drying is negligible. Further, the larger the value of $d\theta/ds$, the greater is the volume of pores holding water within the size class defined by δr ($\alpha 1/\delta s$); so that the value of s when $d\theta/ds$ is at a maximum defines the most frequent pore-size class, as illustrated by curve B in Figure 6.12. The example chosen is typical of a well structured soil showing an adequate volume of large pores between peds, which promote drainage and aeration, relative to

fine pores within peds which hold water in the so-called available range (see below).

Hysteresis

The shape of the moisture characteristic curve depends on whether the soil is drying or wetting—a phenomenon known as *hysteresis*, which occurs for two main reasons:

(a) At low suctions, the large pores do not empty at the same suction as they fill if access to the pore is restricted by narrow necks. Take, for example, a large pore of radius R which is connected to two smaller pores of radii r_1 and r_2 (Figure 6.13a). On drying, the

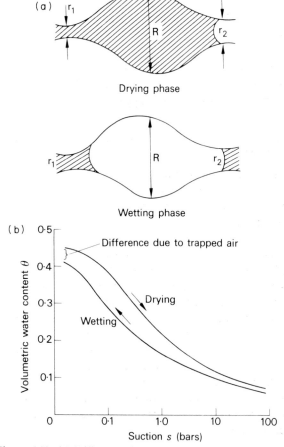

Figure 6.13. (a) Differences in pore water content at the same suction during wetting and drying. (b) Hysteresis in the soil moisture characteristic curve.

large pore will not empty until the suction is great enough to break the meniscus across pore r_1. At this same suction during wetting, the water will advance to fill the pores r_1 and r_2, but the large pore will remain empty until the suction is low enough to maintain an air–water interface across R. Complete filling of the large pore will also be resisted until the pressure of the trapped air is released. The combined effect is that the soil comes to equilibrium at a higher water content for a given suction on the drying than the wetting phase of the cycle, as shown in Figure 6.13b.

(b) The second cause of hysteresis applies mainly to clays which hold appreciable water at high suctions. It is thought to be due to changes in the size and arrangement of the smallest pores as water films contract and clay domains become re-oriented. Such changes may not be reversible.

Available water capacity

The term available water capacity (AWC) was introduced to define the amount of water in a soil that is available for plant growth; the upper limit is set by the field capacity (FC) and the lower limit as the value of θ at which plants lose turgor and wilt, that is, the *wilting point* (WP). When related to the moisture characteristic curve, these water contents are found to correspond to suctions of around 0.05 bars for FC and 15 bars for WP for a number of soil–plant associations.

It is sometimes assumed that water held between FC and WP is equally available for growth, in the sense that a plant will transpire at a constant rate over the whole range. This is approximately true provided that:

(a) The soil moisture characteristic curve is of a type that the bulk of the available water is held at low suctions for which the hydraulic conductivity of the soil is relatively high; that is true of many sandy and sandy loam soils but not of clays.

(b) The evaporative power of the atmosphere is low so that the rate of water movement from soil to root can satisfy the plant's transpiration. This is clearly demonstrated by Figure 6.14 which shows that when the evaporative power of the atmosphere is low (curve C), almost all the available water is transpired before the transpiration rate falls and the plant begins to suffer

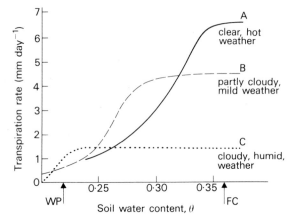

Figure 6.14. The effect of atmospheric conditions and soil water content on plant transpiration rate (after Denmead and Shaw, 1962).

a water deficit. By contrast, when the evaporative power of the atmosphere is very high (curve A), the transpiration rate drops markedly after only a small fraction of the AWC has been consumed.

Nevertheless, the AWC provides a very useful measure of the reserves of soil water, which is relevant to assessing the growth prospects of crops under dryland conditions or to planning the frequency of application of irrigation water (section 13.2).

6.6 EVAPORATION AND EVAPOTRANSPIRATION

The energy balance

Evaporation from open water surfaces E_o is measured using evaporation pans, which are shallow tanks of water exposed to wind and sun (Figure 6.15a); evapotranspiration ET is measured using weighing lysimeters, which are large tanks filled with soil and vegetation (Figure 6.15b). Where such direct measurements are not possible, and especially for the estimation of ET losses on a regional scale, the *potential* rate of evaporation from any surface—open water, plant or wet soil—can be calculated from:

(a) the input of energy into the system and

(b) the capacity of the air to take up and remove water vapour.

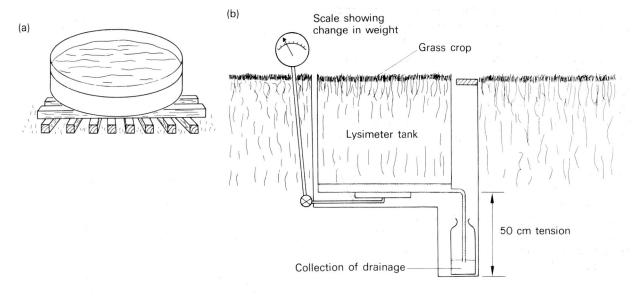

Figure 6.15. (a) Class A evaporation pan. **(b)** A weighing lysimeter (after Pereira, 1973).

The main source of energy is solar radiation of which a fraction is directly reflected, depending on the *albedo* of the surface (ranging from 0.15 for forests to 0.30 for agricultural crops), and some is lost as a net efflux of long-wave radiation from the surface to the atmosphere. Of the energy absorbed H, part is dissipated by evaporation (component E_h), part by the transfer of sensible heat to the air (Q_h) and part stored as heat in the receiving system. Over a 24 hour period, the last component is negligible except for aquatic ecosystems, so we may write:

$$H = E_h + Q_h \qquad (6.13)$$

The transport capacity of the air is estimated from the vapour pressure gradient over the surface and the average wind speed, with an added empirical factor to account for surface roughness when evaporation from vegetation rather than a free water surface is considered.

By making the assumption that in humid temperate climates, Q_h is small so that $H \simeq E_h$, Penman (1970) proposed an approximate formula for the calculation of the daily evaporation rate from a free water surface, or a short green grass crop that completely shades soil in which moisture is not limiting:

$$E = \frac{0.4 R_1}{L} \qquad (6.14)$$

where 0.4 is the fraction of the incoming radiation R_1 which is dissipated by evaporation of water of latent heat L. Inserting $R_1 \simeq 219$ J m^{-2} s^{-1} and $L = 2.4 \times 10^9$ J m^{-3} gives a value of 3 mm day^{-1} for E, which is reasonable if compared with the actual transpiration rates recorded under partly cloudy and overcast conditions, as shown in Figure 6.14. On the other hand, the agreement between actual and potential evapotranspiration rates is likely to break down under high energy inputs, as occur in arid areas, because the rate of supply of water from the soil soon becomes limiting.

Evaporation from soil

STAGES OF DRYING
While the soil surface remains wet, the rate of evaporation is determined by the energy balance between the soil and the atmosphere and the transport capacity of the air. Depending on the albedo of the soil, the rate

can be as high as from an open water surface. This stage is called 'constant rate' drying.

With high evaporation rates from a soil unaffected by a water table, the rate of water loss soon exceeds the rate of supply from below and the surface 1–2 cm become air-dry. Since the macropores empty of water first, the unsaturated permeability for water flow across the dry zone from below decreases sharply and more than offsets the steep rise in suction gradient. The rate of evaporation is now controlled by soil properties and decreases with time, a stage known as 'falling rate' drying. The dry surface layer has a *self-mulching* effect because it helps to conserve moisture at depth, especially under conditions of high insolation in soils with a fine granular or crumb structure.

Water movement in dry soil

Despite the self-mulching effect of the air-dry surface, if evaporation is prolonged and uninterrupted by rainfall, the zone of dry soil gradually extends downwards. This is because there is a slow flow of liquid water through soil wetter than the wilting point, to the drying surface, and there is also some diffusion of water vapour.

The flow of liquid water through relatively dry soil can be demonstrated by the simultaneous transfer of water and its dissolved salts in soil subjected to con-

trolled evaporative conditions. As Figure 6.16 illustrates, the movement of water and chloride from the lower half of a series of soil blocks at initially uniform water contents was identical down to a water content only slightly above the wilting point. Below that it was mainly water that moved, by vapour diffusion, which occurs in response to vapour pressure gradients created by differences in matric suction, salt concentration or temperature. The effect of matric or osmotic suction on vapour pressure is normally small because over the suction range 0–15 bars, the relative vapour pressure merely changes from 1.0 to 0.989. Only when the soil becomes air-dry and the relative vapour pressure of the soil water drops below 0.85 does the vapour pressure gradient under isothermal conditions become appreciable. Temperature, however, has a large effect on vapour pressure, producing a three-fold increase between 10 and 30°C.

An exaggerated diurnal variation in temperature can cause redistribution of salt within the top 15 cm or so of a dry soil through the alternation of vapour and liquid phase transfer of water (Figure 6.17). As the surface

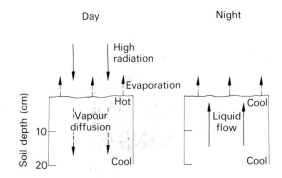

Figure 6.17. Diurnal alternation of vapour and liquid water movement at the surface of a drying soil.

heats up during the day, water vapour diffuses into the cooler layers and condenses. Transfer and condensation of water means heat transfer so that the temperature gradient, and hence the vapour pressure gradient, gradually diminishes until in the evening, with concurrent surface cooling, the temperature differential disappears. At this point, the increased water content and matric potential in the condensation zone induces a return flow of water to the surface, carrying dissolved

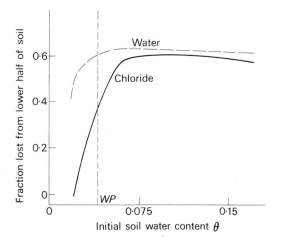

Fig. 6.16. The movement of water and chloride through soil blocks exposed to constant evaporative conditions (after Marshall, 1959).

salts that are deposited when the water subsequently evaporates.

CAPILLARY RISE

Soil water above the water table is drawn upwards in continuous pores until the suction gradient acting upwards is balanced by the gravitational potential gradient. This rise of groundwater, known as *capillary rise*, depends very much on the pore size distribution of the soil, being usually not greater than 1 m in sandy soils but as much as 2 m in some silt loams. Thus, a water table within 1–2 m of the soil surface can lead to excessive evaporative losses, the actual rate of loss being governed by the external atmospheric conditions up to the point where the permeability of the soil becomes limiting. Accumulation of salts at the surface, as occurs in mismanaged irrigated soils, encourages capillary rise because the high osmotic suction maintains an upward head gradient even in moist soils, when the unsaturated permeability is high and appreciable water movement can therefore occur. The rise of groundwater often exacerbates the problem of soil salinity, discussed in Chapter 13.

6.7 SUMMARY

The interaction between soil water and organo-mineral surfaces, exchangeable cations and dissolved salts reduces its partial molar free energy (or chemical potential μ_w) relative to pure water outside the soil, at the same height, temperature and pressure. The *total water potential* ψ is therefore defined as

$$\mu_w \text{ (soil)} - \mu_w^0 \text{ (standard state)} = RT\ln e/e^\circ$$

where e/e° is the relative vapour pressure of the soil water. The value of ψ may be apportioned between several component potentials attributable to matric, osmotic, pressure and gravitational forces, respectively:

$$\psi = \psi_m + \psi_s + \psi_p + \psi_g$$

ψ is usually measured as a pressure p (in bars or Nm^{-2}) which converts directly to *hydraulic head*, according to the equation:

$$\text{Head, } h = p/\rho g$$

Broadly, the interchange of water between the atmosphere and Earth's surface by precipitation P and evaporation E from soil, water and plants comprises the *hydrologic cycle*. Within a defined catchment, $P = E + \text{Streamflow} \pm \text{Deep percolation} \pm \text{Change in soil storage}$, ΔS. Over a year or longer, ΔS is negligible so that streamflow and deep percolation determine the potential water storage, either in surface dams or as groundwater. Nevertheless, changes in ΔS are crucial in the short-term and may decide crop success or failure.

Knowledge of how water penetrates the soil (*infiltration*), moves through the soil (by *saturated flow* when all the conducting pores are filled or *unsaturated flow* at lower water contents) and is lost by evaporation and transpiration (*evapotranspiration*), is therefore vital to the management of water supplies for plant growth and conservation. Rate of water flow v is described by Darcy's law:

$$v = -k \, grad \, h$$

where k is either the saturated or unsaturated permeability. Under static equilibrium conditions, the relationship between the water content θ and soil suction s defines the *moisture characteristic curve*. The available water capacity (*AWC*) amounts to the water held between two defineable points on this curve—the *field capacity* (*FC*), to which most soils drain under gravity, and the point at which plants wilt, the *wilting point* (*WP*).

REFERENCES

CHILDS E.C. (1969) *An Introduction to the Physical Basis of Soil Water Phenomena*. Wiley, New York.

HILLEL D. (1971) *Soil and Water*. Academic Press, New York.

PENMAN H.L. (1970) The water cycle. *Scientific American* **223**, 98–108.

FURTHER READING

BAVER L.D., GARDNER W.H. & GARDNER W.R. (1972) *Soil Physics*. Wiley, New York.

MARSHALL T.J. (1959) Relations between water and soil. *Commonwealth Bureau of Soils, Technical Communication No 50*.

PENMAN H.L. (1963) Vegetation and hydrology. *Commonwealth Bureau of Soils, Technical Communication No. 53*.

PEREIRA H.C. (1973) *Land Use and Water Resources in Temperate and Tropical Climates*. Cambridge University Press.

YOUNGS E.G. & THOMASSEN A. (1975) Water movement in soil, in *Soil Physical Conditions and Crop Production. MAFF Bulletin No. 29*, pp. 228–239.

Chapter 7
Reactions at Surfaces

7.1 ELECTRIC CHARGES AT PARTICLE SURFACES

The inorganic and organic materials which make up the porous structural framework of the soil possess residual electric charges—the *clay minerals* because of internal substitution of elements of the same co-ordination number but different valency, and *humic compounds* because of the dissociation of carboxyl and phenolic groups. In addition, the atoms within any crystal plane of a soil mineral have unneutralized valencies where the interatomic bonding is incomplete at the broken crystal edges.

The net charge arising from isomorphous substitutions within the clay mineral lattice is *negative* and *permanent*, since it is not affected by the concentration and type of ions in the soil solution, unless the pH is so low as to induce decomposition of the crystalline lattice. The result is a mineral with constant surface charge density. On the other hand, the dissociation of carboxyl and phenolic groups and the development of charges on oxygens and hydroxyls at clay or oxide crystal edges, depend upon the solution concentration of protons or hydroxyls which can be reversibly adsorbed at such sites. This charge is therefore *pH-dependent* and H^+ and OH^- are called potential-determining ions, in the same way that Ag^+ and I^- are potential-determining ions at the surface of colloidal silver iodide crystals. Such surfaces have a variable surface charge density.

Surfaces of constant charge

UNIFORM VOLUME CHARGE
In one group of aluminosilicates called the *zeolites*, the permanent negative charge is uniformly distributed over crystal planes within the mineral lattice. Ions dissolved in the solution permeating the mineral can diffuse through a network of interconnecting pores, under-

going frequent changes in electrical potential as they move from the influence of one lattice charge to another. However, when equilibrium is reached between the exchanger and the bathing salt solution, each ion suffers an *average* change in electrical potential on passing into or out of the exchanger, the value of which measures the *Donnan* membrane potential or simply the *Donnan potential*. Cations tend to be retained, or positively adsorbed, within the exchanger and anions are excluded or negatively adsorbed. The volume from which anions are effectively excluded is called the *Donnan free space* (Figure 7.1). This model applies

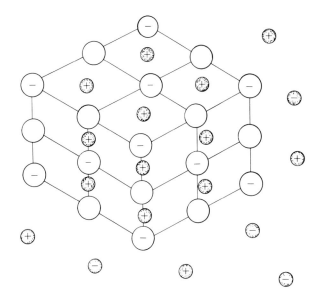

Figure 7.1. Concentration of mobile cations ($\oplus$) within the Donnan Free Space and exclusion of mobile anions $\ominus$).

equally well to soil humic polymers and plant constituents—cell walls and cytoplasmic proteins, except that the charges on these are not permanent but pH-dependent.

NON-UNIFORM VOLUME CHARGE

Unlike the zeolites, clay minerals have thin laminar structures in which the permanent negative charge acts as if it were smeared out over the planar surfaces of the crystals. As before, the negative charge creates a surplus of cations (the counterions) and deficit of anions (the co-ions) in the solution immediately adjacent to the surface. The simplest case is one in which each surface charge is neutralized by the close proximity of a mobile charge of opposite sign—the parallel alignment of charges in two planes is called a *Helmholtz double layer* (Figure 7.2).

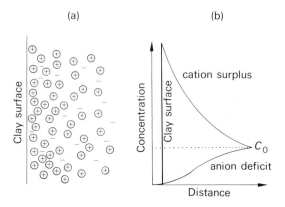

Figure 7.3. (a, b) The ionic distribution at a negatively charged clay surface.

CHARACTERISTICS OF THE GOUY DOUBLE LAYER

The concentration profiles for the cations and anions of a single salt solution at a negatively charged surface are illustrated in Figure 7.3b, where C_0 represents the concentration of the solution away from the influence of the charged surface. The cation concentration attains a very high value at the surface, while the anion concentration becomes vanishingly small. At a clay mineral planar surface, the volume density of charge in the diffuse layer must remain constant irrespective of changes in the ionic composition of the solution. For each square cm of surface, part of the charge (Γ) is neutralized by the attraction of cations and part by the repulsion of anions so that

$$\Gamma = \Gamma_+ + \Gamma_- \qquad (7.1)$$

where Γ_+ is the cationic charge in the diffuse layer (me cm^{-2}) and Γ_- is the anionic charge repelled from the diffuse layer (me cm^{-2}). Schofield (1947b) showed that provided the surface charge density was $\geq 1 \times 10^{-7}$ me cm^{-2}, the quantity of Cl$^-$ ions repelled was given by the equation:

$$\frac{\Gamma_-}{C_0} = -\frac{q}{\sqrt{v\beta C_0}} \qquad (7.2)$$

where Γ_- = me cm^{-2} of repelled Cl$^-$ ions

C_0 = concentration of the original solution (me cm^{-3})

v = valency of the cation

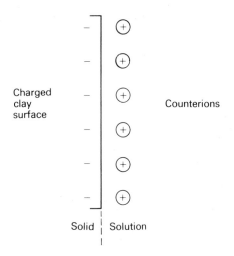

Figure 7.2. The Helmholtz double layer.

But in aqueous solution the high permittivity of water reduces the Coulombic force of attraction between the fixed and mobile charges, which is also opposed by a diffusive force tending to drive the cations out of the region of high concentration near the charged surface. At equilibrium, the Coulombic force across a plane of unit area near to and parallel to the surface is just balanced by the difference in osmotic pressure between the plane and the external solution far removed from the surface. The result is a diffuse distribution of cations and anions in solution, which together with the surface charge comprises the *Gouy double layer*, so named after the man who first described it mathematically (Figure 7.3a).

β is a constant (of value 1.06×10^{15} cm me^{-1} for water at 25 °C) and q is a term dependent on the valency ratio of the attracted and repelled ions ($q = 2$ for NaCl and 1.46 for CaCl$_2$).

The term $q/\sqrt{v\beta C_0}$ may be visualized as the volume of solution per unit area of surface (i.e. a distance) from which Cl$^-$ ions would be totally excluded, and is referred to as the *effective thickness* of the diffuse layer (d). Calculated values of d for different concentrations of NaCl and CaCl$_2$ solutions in contact with a clay surface are given in Table 7.1, from which it is

Table 7.1. Calculated diffuse layer thicknesses for different concentrations of 1 : 1 and 2 : 1 electrolytes.

Electrolyte concentration, C_0	Effective diffuse layer thickness, d	
	NaCl	CaCl$_2$
(me cm^{-3})*	(nm)	
0.1	1.94	1.0
0.01	6.2	3.2
0.001	19.4	10.1

* A solution containing 1 me of salt per cm^3 is defined as a *normal* (N) solution.

seen that d is very nearly proportional to $1/v$ and is exactly proportional to $1/\sqrt{C_0}$. In short, as the valency of the cation increases and the solution concentration increases, the diffuse layer is compressed.

MODIFICATIONS TO THE GOUY DOUBLE LAYER THEORY
The Gouy theory works well for monovalent cation solutions of intermediate concentrations (0.1–0.0001 N) and surface charge densities of $1–4 \times 10^{-7}$ me cm^{-2}. It breaks down once certain limits of cation valency and solution concentration are reached—firstly because the ions are assumed to be point charges, which leads to absurdly high concentrations of cations in the inner region of the diffuse layer at surfaces of high charge density; secondly because the theory ignores the existence of forces between the surface and counterions other than simple Coulombic forces. It predicts that cations of the same valency will be adsorbed with the same energy, whereas it is known from cation exchange

measurements that Li$^+$ and K$^+$, or Ca^{2+} and Sr^{2+} for example, are not held with the same tenacity by clays in contact with solutions of identical normality. These factors are accommodated in the *Stern model*—essentially a combination of the Helmholtz and Gouy concepts—which splits the solution component of the double layer into two parts:
(a) A plane of cations of finite size located within a few Angstroms* of the surface—the Stern layer, across which the electrical potential decays linearly.
(b) A diffuse layer of cations, across which the potential falls almost exponentially, the inner surface of which abuts onto the Stern layer and is called the outer Helmholtz plane (*OHP*). (Figure 7.4). Allowance is

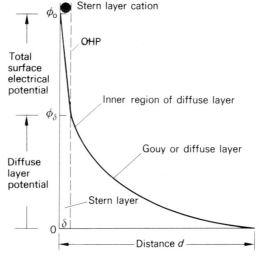

Figure 7.4. Electrical potential gradients at a planar clay surface (after van Olphen, 1977).

made in the Stern model for the reduction in permittivity of the water due to the high field strength very near the surface, and for specific adsorption forces between the surface and Stern layer cations.

SPECIFIC ADSORPTION OF CATIONS
At a distance from the surface < 4Å the assumption of a plane of uniform surface charge does not hold and cations in the Stern layer experience a polarization

* 1 Angstrom (Å) $= 10^{-10}$ m

force in the vicinity of each charged site. Depending on the balance of forces—the polarizing effect of the cation on the water molecules in its hydration shell compared to the polarization of the cation by the surface charge—the cation may shed its water of hydration to enter the Stern layer and reside directly on the surface where it forms an *ion-pair* with the charged site. The group of elements Li, Na, K, Rb, Cs forms a lyotropic series in which the hydration energy of the monovalent cation decreases with increasing atomic size; the small cation Li^+ tends to remain hydrated at the surface whereas the large Cs^+ ion tends to dehydrate and become tightly adsorbed. Cations held in the Stern layer by specific adsorption forces are continually exchanging with cations in the diffuse layer, but at any instant the proportion held in the Stern layer increases from about 16 per cent for Li^+, to 36 per cent for Na^+ and 49 per cent for K^+, and the overall strength of adsorption increases in the sequence $Li < Na < K < Rb < Cs$. By the same reasoning,

there is an increase in the strength of adsorption of the divalent cations in the order $Mg < Ca < Sr < Ba < Ra$.

Taking the joint Stern–Gouy layer theory into account (Figure 7.4), one finds that an increase in solution concentration not only compresses the double layer but also increases the proportion of ions in the Stern layer. The diffuse layer potential ϕ_δ decreases relative to the total surface potential ϕ_0 and the repulsion of anions from the inner region of the diffuse layer is diminished.

Surfaces of variable charge

OXIDE SURFACES

Free oxides in soil, especially the hydrated Fe and Al oxides, exhibit pH-dependent charges due to the reversible adsorption of potential-determining H^+ ions. The change in surface charge with pH change is illustrated for hydrated Al_2O_3 in Figure 7.5. The pH at which the net charge on the surface is zero defines the

Figure 7.5. Changes in surface charge on an aluminium oxide.

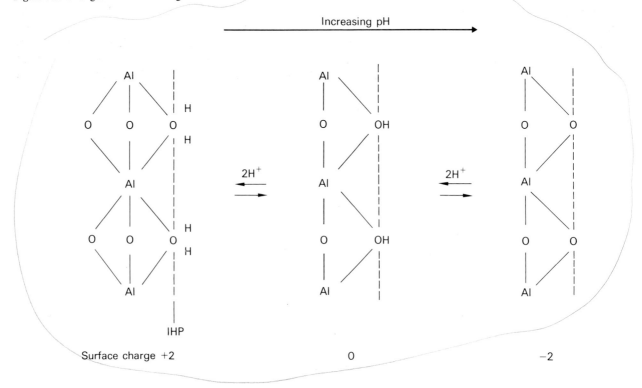

point of zero charge (PZC). The greater the polarizing effect of the metal atom on the O–H bond, the more acidic is the surface; that is, the lower is the *PZC*. Silica, for example, has a *PZC* <2 whereas the *PZC* of ferric oxide is ~ 9.

The oxygen, OH and OH_2^+ groups of the oxide surface lie in the inner Helmholtz plane (*IHP*) (Figure 7.5); naturally, when the surface has a net positive charge, anions are attracted from the solution to form a double layer. If the charge on the surface is all balanced by counterion charge in the Stern layer (between the surface and the *OHP*), so that the diffuse layer potential is zero, then the solution pH defines the *isoelectric point (IEP)* of the solid.

CLAY MINERAL EDGE FACES

Aluminium atoms coordinated to O and OH groups are also exposed at the edges of clay crystals. This structure differs from that of the free oxides only in that the oxygens are also coordinated to Si atoms in at least one contiguous silica layer. The full negative charge of 2 electrons per 0.4 nm² of edge area is only developed above pH 9 when the $\equiv Si–OH$ groups dissociate (see Figure 2.16). The association of H^+ ions with the O and OH coordinated to the aluminium increases as the pH decreases from 9 to 5 in the manner illustrated in Figure 7.6, the *PZC* of the edge face

occurring at *c.* pH 7 for kaolinite. Thus, the edge faces of the clay minerals show a pH-dependent charge.

ORGANIC MATTER

Although the charge of humified organic matter is completely dependent on pH due to the dissociation of carboxyl and phenolic groups (section 3.4), the charge is negative from pH 3 upwards and so augments the permanent negative charge of the clay minerals. In soils of high organic content, however, it has the effect of considerably increasing the pH-dependence of the soil's *CEC*.

The significance of pH-dependent charges in soil

Edge charges are of greatest significance in the kaolinites which have low permanent charges (2–5 me per 100 g) and high edge:planar area ratios (1:10 to 1:5). The development of one positive charge per 0.4 nm² of edge area amounts to 0.004 me m⁻², or 0.4 to 4 me per 100 g of kaolinite ranging in edge face area from 1 to 10 m² g⁻¹. Conversely, edge charges are unimportant in the 2:1 clay minerals which have high permanent negative charges (40–150 me per 100 g) and exist as smaller crystals, especially in the direction of the *C* axis, thereby offering a smaller edge:planar ratio than the kaolinites. The charge characteristics of soil clays may, however, be considerably modified from that of the

Figure 7.6. Charge development at the edge face of kaolinite with change in pH.

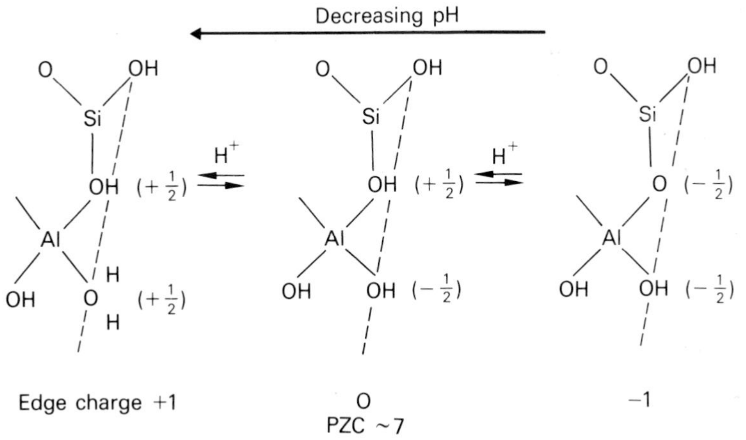

pure clay minerals by the presence of sesquioxides in variable amounts and spatial distribution.

The formation and occurrence of iron and aluminium oxides was discussed briefly in section 2.4. With *PZC* values in the range 7–9, they are normally positively charged in soil and occur as thin films adsorbed on the planar surfaces of clays, as illustrated by the $Fe(OH)_3$ deposit on kaolinite in Figure 7.7; $Al(OH)_3$ films occur

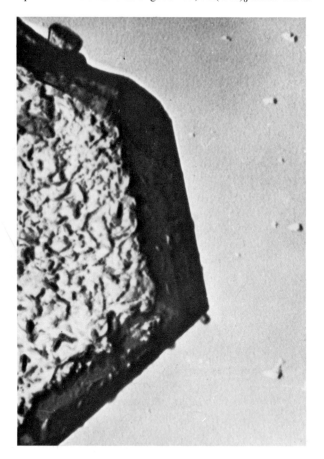

Figure 7.7. $Fe(OH)_3$ deposit on a kaolinite cleavage face (courtesy of D.J. Greenland).

preferentially within the interlamellar spaces of micaceous-type clay minerals to form mixed layer minerals. Positive charge densities of 20 to 80 me per 100 g have been recorded for soil oxides, so that soils which are low in organic matter and contain mainly kaolinite and sesquioxides can develop a *net positive* charge at low pH. Such behaviour is prevalent in highly weathered soils of the Tropics, exemplified by an acid subsoil from South Africa, the charge *vs* pH curve of which is illustrated in Figure 7.8. The presence of significant

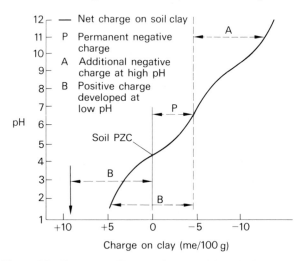

Figure 7.8. Charge *vs* pH curve for an acid tropical red loam (after Schofield, 1939).

positive charge densities, wholly dependent on pH, has profound implications not only for the ability of the soil to retain cations, but also for the adsorption of anions, two topics which are discussed in the following sections.

7.2 CATION EXCHANGE

Exchangeable cations

A high proportion of the cations released by rock weathering and organic decomposition are adsorbed by the clay and humus colloids, leaving a low concentration in the soil solution to be balanced by mineral anions and bicarbonate generated from the respiration of soil organisms. The cations held in the Stern–Gouy double layer are called *exchangeable*, in that a cation in solution can exchange with a cation in the double layer. In all but the most acid or alkaline of soils, the major cations are Ca^{2+}, Mg^{2+}, K^+, and Na^+, in roughly the proportions of 80% Ca: 15% Mg: 5% (Na + K) with variable amounts of NH_4^+, depending on the efficiency of microbial nitrification (section 8.3). Trace amounts of other cations such as Cu^{2+}, Mn^{2+} and Zn^{2+} are also adsorbed, although chelation with organic compounds and precipitation play a more prominent part in the retention of these elements (section 10.5).

Cation exchange capacity and exchange acidity

One method of determining soil *CEC*, as defined in section 2.5, is to displace all the exchangeable cations with NH_4^+ ions provided by a 1N solution of ammonium acetate at pH 7, and to measure the amount of NH_3 which can be distilled off by steam at pH >10. The pH of the displacing solution is specified because of the pH-dependence of the *CEC*. In calcareous soils the sum of the cations ($\sum$Ca, Mg, K, Na) is invariably equal to the *CEC* since any deficit of the cations on the exchanger can be made up by Ca^{2+} ions from the dissolution of $CaCO_3$; in non-calcareous soils, however, $\sum$Ca, Mg, K, Na is frequently less than the *CEC*, the difference being referred to as the *exchange acidity* (in me H^+ per 100 g soil). The ratio:

$$\frac{\sum(\text{Ca, Mg, K, Na})}{CEC} \times 100$$

defines the *per cent base saturation*, which is correlated with the pH in many mildly acid and neutral soils.

The exchangeable acidity is not all due to H^+ ions. As the soil pH drops below 5, increasing amounts of Al^{3+} ions can be displaced by leaching with strong salt

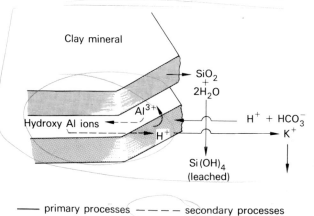

— primary processes − − − − secondary processes

Figure 7.9. Clay mineral weathering in an acidic environment.

solutions. Under acid conditions the weathering of clay minerals is accelerated, with the release of Al, SiO_2 and smaller amounts of Mg, K, Fe and Mn (Figure 7.9). While the silicic acid formed is removed by leaching the Al, Mg, K and Mn are retained initially as exchangeable cations. Iron (Fe^{3+}) salts are very insoluble except at pH values <3, so that the ferric Fe is rapidly immobilized as the insoluble oxide, carbonate or phosphate. With the exception of illitic clay soils, K^+ is also lost by leaching so that an acid clay remains which is dominated by Al^{3+} with some Mg^{2+} (and H^+ ions produced by the hydrolysis of the Al^{3+}).

pH buffering capacity

High levels of exchangeable aluminium greatly increase the soil's capacity to neutralize OH^- ions, which is a measure of the soil's *pH buffering capacity*. The hydrated Al^{3+} ion hydrolyses on the addition of alkali according to the reaction:

$$(Al6H_2O)^{3+} + H_2O \rightleftarrows [Al(OH)\,5H_2O]^{2+} + H_3O^+ \quad (7.3)$$

there being equal concentrations of the di- and trivalent species at pH 5. The hydroxyaluminium ions $(AlOH)^{2+}$ show a pronounced tendency to combine through bridging OH groups and build up into polymeric units of 6 and more Al atoms. The consequent release of H^+ contributes to the buffering capacity until eventually neutralization is complete and amorphous $Al(OH)_3$ precipitates on the surfaces and interlamellar spaces of the clay.

Three stages are recognized in the neutralization of acid clays and these are illustrated in Figure 7.10 where representative curves for kaolinite, vermiculite and montmorillonite titrated in a neutral salt solution are plotted. Provided that sufficient time is allowed for near-equilibrium pH values to be attained, the three clays show markedly different buffering within the ranges:

I $<$ pH 4 exchangeable H^+
II pH 4–5.5 exchangeable Al^{3+}
III pH 5.5–7.5 hydroxyaluminium ions

The quantity of alkali added to attain a chosen higher pH is a measure of the *titratable acidity* (in me per 100 g), and in practice can be used to estimate the soil's

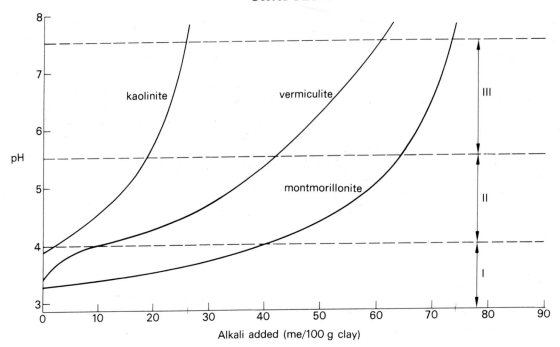

Figure 7.10. pH-titration curves for kaolinite, vermiculite and montmorillonite in 0.1N KCl.

lime requirement (section 11.3). In soils with much interlayer and surface hydroxy-Al films, the titratable acidity is always greater than the exchangeable acidity because part of the *CEC* is blocked by hydroxy-Al deposits which are not displaced by leaching with neutral salt solutions.

Cation exchange reactions

EXCHANGE EQUATIONS
The technique of using a strong solution of NH_4^+ ions to displace the exchangeable cations and determine the *CEC* is an example of *cation exchange*. Cation exchange is a reversible process in which one equivalent of a cation in solution replaces one equivalent of cation on the exchanger, as for example:

$$(NH_4^+) + (Ca\text{-exchanger}) \rightleftarrows (NH_4\text{-exchanger}) + \tfrac{1}{2}(Ca^{2+})$$
(7.4)

where (NH_4) and (Ca^{2+}) are the molar activities of the ions in solution, and (Ca-exchanger) and (NH_4-

exchanger) represent the activities of the ions on the exchanger. Much has been written about the validity of cation exchange equations, which all founder on the problem of measuring the *activity* of ions in the adsorbed state. The best approach is probably an empirical one in which it is assumed that NH_4^+ and Ca^{2+} are adsorbed in proportion to (a) the 'reduced' ratio of their equilibrium activities in solution and (b) the relative affinity of the exchanger for NH_4^+ compared to Ca^{2+}, as measured by a selectivity coefficient k_{NH_4-Ca}. The resultant equation:

$$\frac{NH_4\text{-exchanger}}{Ca\text{-exchanger}} = k_{NH_4-Ca}\frac{(NH_4^+)}{(Ca^{2+})^{\frac{1}{2}}}$$
(7.5)

is called the *Gapon equation* and has proved of wide applicability to cation exchange in soils. Provided that the monovalent cation (Na^+ or K^+) amounts to < 50 per cent of the exchangeable Ca^{2+} and Mg^{2+} combined, reasonably constant values of the Gapon coefficients $k_{Na-Ca, Mg}$ and $k_{K-Ca, Mg}$ are obtained for soils of similar clay mineral type from the equations:

$$\frac{\text{Na-exchanger}}{\text{(Ca + Mg)-exchanger}} = k_{\text{Na-Ca, Mg}} \frac{(\text{Na}^+)}{(\text{Ca}^{2+} + \text{Mg}^{2+})^{\frac{1}{2}}} \quad (7.6)$$

and

$$\frac{\text{K-exchanger}}{\text{(Ca + Mg)-exchanger}} = k_{\text{K-Ca, Mg}} \frac{(\text{K}^+)}{(\text{Ca}^{2+} + \text{Mg}^{2+})^{\frac{1}{2}}} \quad (7.7)$$

(Ca^{2+} and Mg^{2+} have sufficiently similar adsorption affinities, relative to the monovalent cations, to be regarded as interchangeable). Equation (7.6) has been of great value in predicting the likely 'sodium hazard' of irrigation waters (section 13.2), and equation (7.7) has been applied to the study of potassium availability in soil.

For example, equation (7.7) may be re-written as

$$\frac{\text{exchangeable K}}{\text{CEC-exchangeable K}} = k' \frac{\text{K}}{\sqrt{\text{Ca} + \text{Mg}}} \quad (7.8)$$

where the quantities on the left hand side of the equation are in me per 100 g soil. When exchangeable K is small (<10 per cent of the CEC), equation (7.8) reduces to the simple form

$$\text{exchangeable K} = k'' \frac{\text{K}}{\sqrt{\text{Ca} + \text{Mg}}} \quad (7.9)$$

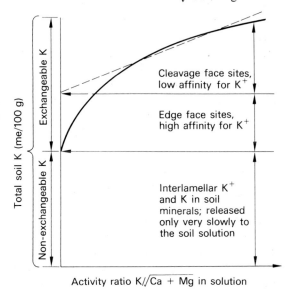

Figure 7.11. Schematic Q/I relation for K in a micaceous clay soil (after Beckett, 1971).

where k'' includes the Gapon constant and the CEC of soil. Usually the *change* in exchangeable K (ΔK) is plotted against the activity ratio (AR) to give a *quantity/intensity* relation (Q/I) for the soil's exchangeable potassium (Figure 7.11). However, the reactions of K in soil are complex, especially in soils containing much partially-expanded clay mineral. As described in section 2.4, K^+ ions in the unexpanded interlamellar spaces of illite and mica are non-exchangeable, but with weathering and the depletion of K in solution, the clay crystals slowly exfoliate from the edges exposing the interlamellar K to exchange by other cations in solution (Figure 7.12). The exposed interlamellar sites at the

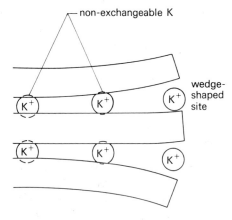

Figure 7.12. The weathering edge of a micaceous clay crystal.

crystal edges (which have been likened to 'wedge-shaped' sites) have a higher affinity for K^+ than the cleavage faces so that the whole crystal exhibits a changing affinity for K^+ relative to Ca^{2+} and Mg^{2+}, as first the edge sites and then the cleavage face sites are occupied. This is reflected in a gradual change in the slope of the Q/I plot, and hence the soil's buffering capacity for K, in going from low to high AR values.

THE RATIO LAW

The Gapon equation indicates that if the proportions of Na^+ and K^+ on the exchanger are unchanged, the ratio Na^+/K^+ in the equilibrium solution remains constant, and similarly, if the proportions of K^+ and Ca^{2+} on the exchanger are constant, the ratio $\text{K}^+/\sqrt{\text{Ca}^{2+}}$ is also constant. Schofield (1947a) was the first to formulate such observations into a general hypothesis which he called the *ratio law*. It states that 'when cations in solution are in equilibrium with a larger number of exchangeable cations, a change in the concentration of the solution will not disturb the equilibrium if the concentrations of all the monovalent ions are changed in the one ratio, those of all the divalent ions in the square of that ratio and those of all the trivalent ions in the cube of that ratio.' For 'concentration' one may read 'activity'.

The constancy of the cation activity ratios depends on the effective exclusion of the accompanying anions such as Cl^- from the adsorption sites, since only then will the activities of the exchangeable cations be unaffected by changes in the Cl^- concentration of the external solution. Chloride exclusion is determined by the thickness of the double layer (section 7.1) so that the ratio law is only valid provided that

(1) There is a preponderance (> 80 per cent) of negative over positive charge on the surface, and

(2) The total solution concentration is not too high. As theory predicts, the upper limit of concentration for which the ratio law applies varies with the valency of the cation, being approximately 0.1 N for Na and K, 0.02 N for Ca and 0.005 N for Al. Despite these limitations, the ratio law is very useful in interpreting ion exchange in soils as the following examples illustrate.

Leaching of cations

Percolation of water through the soil leads to losses of solutes by leaching. If the concentration of cations in the leaching solution is adjusted according to the ratio law, the equilibrium between exchangeable and solution cations will be undisturbed despite the overall reduction in concentration. This is confirmed by the results of percolation tests on an unlimed Rothamsted Park Grass soil (Table 7.2) in which the concentrations of cations in the second leaching solution were reduced in accordance with the ratio law. With percolating rain water, which is naturally very dilute, all the cations are in low concentration, and exchange occurs between cations of higher valency in solution and adsorbed cations of lower valency so that the activity ratios $Na/\sqrt{Ca}$, $K/\sqrt{Ca}$ and so on tend to remain constant. Gradually Na followed by K is lost, the exchange

Table 7.2. Rothamsted Park Grass Soil, at equilibrium with solution A and percolated with solution B, the concentration of which was adjusted according to the ratio law.

Solution	Percolate No.	Na^+	K^+	Mg^{2+}	Ca^{2+}	Total cations
			(me l^{-1})			
A	3	0.0	3.9	1.3	5.1	10.3
	(at equilibrium)					
B	1	0.0	2.1	0.3	1.5	3.9
B	2	0.0	2.0	0.3	1.3	3.6
B	3	0.0	2.0	0.4	1.2	3.6
	(at equilibrium)					

(after Schofield, 1947a)

surfaces becoming dominated by Ca and Mg and ultimately by Al. The result is the genesis of an acid soil of low base saturation described earlier.

pH measurement

Soil pH is frequently measured after shaking one part of the soil by weight with 2.5 or 5 parts of distilled water by volume and immersing the tips of a glass and a calomel reference electrode in the supernatant solution. The dilution of the soil solution inevitably means that the measured pH is higher than that of the undisturbed soil, an effect especially noticeable in saline soils. But if the soil is shaken with a solution containing the most abundant cation at a concentration approximating to the total concentration of the soil solution, the measured pH changes little with dilution. For many soils a solution 0.01M $CaCl_2$ is found to satisfy these requirements and the pH so measured is very close to the natural pH of the soil, as determined in a solution which has been equilibrated with successive samples of the soil (Table 7.3).

Table 7.3. pH measurements on a basaltic red loam.

Soil/liquid ratio	pH in H_2O	pH in 0.01M $CaCl_2$	pH in an equilibrium solution*
(g ml^{-1})			
1/2	5.08	4.45	4.45
1/5	5.29	4.45	4.45
1/10	5.43	4.46	4.45
1/50	5.72	4.52	4.45

* 0.01 M $CaCl_2$ after successive equilibrations with 4 separate samples of soil (after White, 1969).

7.3 ANION ADSORPTION

The adsorption sites

When positively charged, the pH-dependent sites on clay minerals and sesquioxides attract anions which may be exchanged by other anions in the soil solution, in accordance with the ratio law—that is, the ratio of adsorbed $H_2PO_4^-$ to $SO_4^=$ varies with the activity ratio $H_2PO_4^-/\sqrt{SO_4^=}$ in the solution provided that the concentration of cations at the surface is vanishingly small. However, when the positive sites are only isolated spots on a clay mineral surface, or the soil solution concentration is low enough for the diffuse layer at planar surfaces to envelope the positive sites at the edges, anion adsorption is determined by the characteristics of the double layer at the negatively charged surfaces (Figure 7.13a). In consequence, the concentration of $SO_4^=$ at the inner surface of the double layer is less than the concentration of $H_2PO_4^-$, and as the soil solution becomes more dilute, the increase in the concentration of $SO_4^=$ in the bulk solution is relatively greater than the

increase in $H_2PO_4^-$ concentration. Under these conditions the cation–anion *activity products* $(Ca^{2+}) \times (SO_4^=)$ and $(Ca^{2+}) \times (H_2PO_4^-)^2$ have constant values.

The situation is further complicated by two other factors:

(a) Specific forces acting between surface Al and Fe atoms and anions in solution;
(b) The formation of surface complexes and the precipitation of insoluble salts.

Specific adsorption of anions

Large anions such as $H_2PO_4^-$, $HSiO_3^-$, HCO_3^- and $MoO_4^=$ are only weakly hydrated, and can approach very closely to a site of positive charge to form ion-pairs in the manner described for the specific adsorption of weakly hydrated cations (section 7.2). But the adsorption of di- and trivalent anions is made more complicated than for simple anions like Cl^- and NO_3^- because the charge on both the oxide surface and the anion varies with pH, and because phosphate, silicate and to a lesser degree sulphate have a chemical

Figure 7.13. (a, b) Physical models of the interrelations between clays, sesquioxides and anions in solution.

(a)

(b)

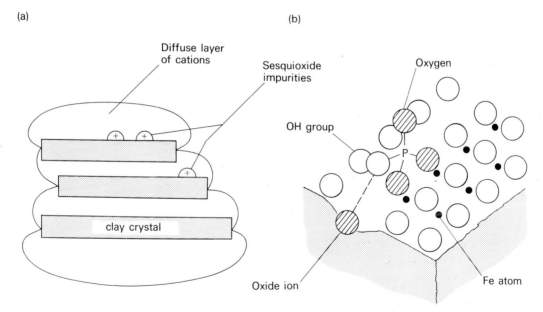

Sites for anion adsorption on clay crystals

Model of $HPO_4^=$ adsorption on a goethite crystal surface after Parfitt and others (1976)

affinity for Fe and Al, as demonstrated by their ability to enter the co-ordination shell of the metal ion. The displacement of an OH or $(OH_2)^+$ group from the co-ordination shell of the metal by an O atom of the $H_2PO_4^-$ or $HPO_4^=$ ion is an example of *ligand exchange*, which leads to chemisorption of the anion on the surface.

Examination of the IR spectra of phosphated iron oxide (goethite) suggests that phosphate displaces two singly co-ordinated OH groups of adjacent Fe atoms to form a binuclear HPO_4 bridging group (Figure 7.13b). Adsorption of the phosphate decreases the potential for protonation of the surface at low pH so that the *PZC* of the oxide is reduced. Phosphate and other anions attracted to positive sites on oxide surfaces can be desorbed by reducing the ion's concentration in the ambient solution, but anions co-ordinated to the surface can only be desorbed by raising the pH or introducing another anion with a higher affinity for the metal of the oxide.

Precipitation of insoluble salts

Phosphate also forms insoluble salts with Fe, Al and Ca. The crystalline minerals *strengite* ($FePO_4 . 2H_2O$) and *variscite* ($AlPO_4 . 2H_2O$) are stable below *c.* pH 1.4 and 3.1 respectively, but in less acid soils amorphous compounds with P:Al mole ratios approaching 1 can form as phosphate in solution reacts with adsorbed hydroxy-aluminium ions (section 10.3). Above pH 6.5, phosphate forms insoluble salts with Ca such as *octacalcium phosphate* $Ca_4H(PO_4)_3 . 5H_2O$, which reverts to the more stable *hydroxyapatite* $Ca_{10}(PO_4)_6(OH)_2$, and phosphate may also be present as an impurity in calcite crystals. In the latter case, the release of phosphate depends on calcite dissolution, which is primarily controlled by the partial pressure of CO_2 in the soil air (section 11.3).

7.4 PARTICLE INTERACTION AND SWELLING

Attractive and repulsive forces

The attractive forces between clay particles are of two main kinds:

(a) *Electrostatic*, as between negatively charged cleavage faces and positively charged edges; or K^+ ions and the adjacent lamellae in an illite crystal; or partially hydrated Ca^{2+} and Mg^{2+} ions and the lamellae of vermiculite;

(b) *Van der Waals' forces* between individual atoms in two particles. For individual pairs of atoms the force falls off with the seventh power of the interatomic distance, but the total force between two particles is the sum of all the interatomic forces which results not only in an appreciable force between macromolecules, but also in a less rapid decay in force strength with distance.

The origin of the attractive forces is such that they vary little with changes in the solution phase, whereas the repulsive forces which develop due to the interaction of diffuse double layers are markedly dependent on the composition and concentration of the soil solution. The model of the double layer presented so far assumes no impediment to the diffuse distribution of ions at the charged surface—the concentration in the midplane between two clay plates is equal to the concentration in the bulk solution. This is reasonably true of dilute clay suspensions in salt solutions, but in concentrated suspensions and in natural soils, the particles are close enough for the diffuse layers to interact, when the balance between forces of attraction and repulsion determines the *net attractive force* between the particles. If the net force is attractive (Figure 7.14a) the particles remain close together and are described as *flocculated*. It also follows that soil aggregates of which the clay particles form a part will also remain stable. Conversely, if the net force is repulsive (Figure 7.14b) the particles move farther apart until a new equilibrium position is attained, under the constraint of an externally applied force, or the particles move far enough to exist as separate entities in the solution—a state described as *deflocculated*.

The excess concentration of ions at the midplane between two interacting particles gives rise to an osmotic or *swelling pressure* of the solution between the two plates. Swelling pressures of 10–20 bars may develop for particle separations from < 1 nm upwards, depending on the type of clay mineral, the interlamellar cations and the salt concentration of the solution.

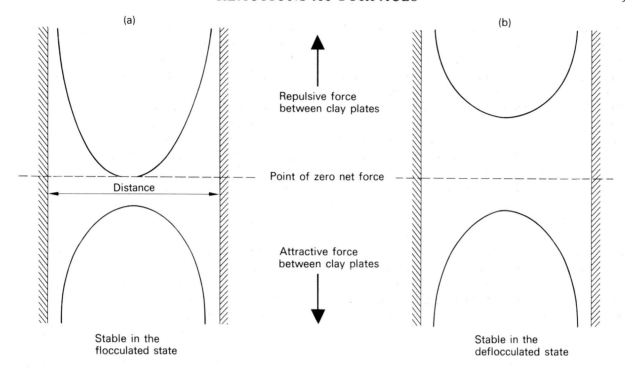

Figure 7.14. (a, b) The interaction of attractive and repulsive forces between clay plates in salt solutions.

Flocculation–deflocculation and swelling behaviour

FACE-TO-FACE FLOCCULATION

Lamellae of the hydrous micas stack with planar surfaces opposed (face-to-face) to form highly oriented crystals wherein K^+ is the dominant interlayer cation. Similar units are formed by the montmorillonites when Ca^{2+} and Mg^{2+} predominate as interlayer cations, although the structures are not as well oriented as in the

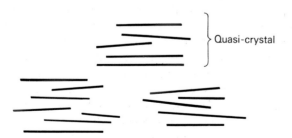

Figure 7.15. Face-to-face flocculation of Ca-saturated montmorillonite.

micas so that the term *quasi-crystal* is preferred (Figure 7.15). Surface areas for such clays of 100–150 m² g⁻¹, compared to 750–800 m² g⁻¹ for completely dispersed montmorillonite, suggest that each crystal consists of 5–8 lamellae. Anions are totally excluded from the interlamellar regions and the diffuse double layer develops only at the external surfaces of the crystal. The interlamellar cations appear to occupy a midplane position and provide a strong attractive force which stabilizes the interlamellar spacing, irrespective of the external solution concentration.

SWELLING BEHAVIOUR

Except in atmospheres drier than about 50 per cent R.H. (section 6.1), Ca-saturated montmorillonite holds at least two layers of water molecules in the inter-lamellar spaces due to the hydration of the cations. The basal spacing is then 1.5 nm. Additional water is taken up as the soil solution becomes more dilute, but only to the extent of expanding the basal spacing to

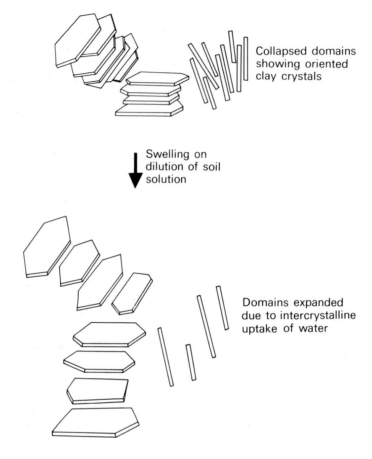

Collapsed domains
showing oriented
clay crystals

Swelling on
dilution of soil
solution

Domains expanded
due to intercrystalline
uptake of water

Figure 7.16. Intercrystalline or domain swelling of a 2 : 1 clay
(after Russell, 1973).

c. 2 nm which is maintained even in distilled water. This behaviour is typical of *intracrystalline* swelling in Ca-saturated 2:1 clays.

As observed in section 4.4, however, crystals of the 2:1 clays may orientate to form units up to 5 μm thick which are called *domains*. Water is adsorbed at the outer surfaces of the crystals within the larger units, leading to swelling and the creation of wedge-shaped spaces and pores where water can be held by surface tension. Swelling of this kind, described as *intercrystalline* or domain swelling, is illustrated in Figure 7.16; it induces particle realignments that release the strains imposed by the crystal packing and bending during the previous drying cycle, and has been suggested as a cause of hysteresis in moisture characteristic curves at high suctions (section 6.5).

The swelling pattern of montmorillonite is entirely changed when divalent cations are replaced by the monovalent cations Na^+ and Li^+ (Table 7.4). As the salt concentration drops below 0.3N NaCl, repulsive forces prevail and the swelling pressure forces the lamellae apart to distances which can be predicted from diffuse layer thicknesses calculated from Gouy theory.

EDGE-TO-FACE FLOCCULATION
Kaolinite crystals are larger than those of montmorillonite and the hydrous micas and they flocculate face-to-face in concentrated solutions at high pH. Leaching

Table 7.4. Intracrystalline swelling of 2 : 1 clays saturated with different cations.

Clay mineral	Charge per unit cell	Exchangeable cation	Swelling in dilute solution
			(nm)
Mica	−2.0	K, Na	none
Vermiculite	−1.3	Ca, Mg, Na	1.4–1.5
		K	1.2
		Cs	1.2
		Li	≫4.0
Montmorillonite	−0.67	Ca, Mg	1.9
		K	1.5 and ≫4.0
		Cs	1.2
		Na, Li	≫4.0

(after Quirk, 1968)

and a reduction in the concentration of the soil solution cause the diffuse layers to expand, and the increased swelling pressure to disrupt the face-to-face arrangement. If there is a concurrent increase in soil acidity, the edge faces become positively charged and the mutual attraction of double layers at cleavage and edge faces encourages *edge-to-face* flocculation (Figure 7.17).

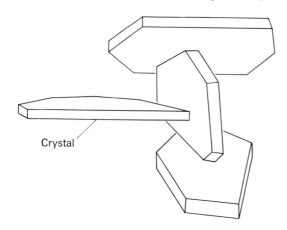

Crystal

Figure 7.17. Edge-to-face flocculation of kaolinite at low pH.

Of the two mechanisms, face-to-face flocculation is the norm for montmorillonitic clays, depending as it does on the predominance of di- and trivalent exchangeable cations or the precipitation of sesquioxidic films which form an 'electrostatic bridge' between opposing surfaces. However, edge-to-face flocculation is more common in kaolinites because of the high positive edge charge. Edge-to-face flocculation is insensitive to changes in the solution concentration but is weakened by the adsorption of anions such as $H_2PO_4^-$ at the edge faces. In the same way, chemisorption of phosphate and organic anions by sesquioxides may actually reverse the surface charge, rendering the oxide negatively charged and ineffective as a clay flocculant.

7.5 CLAY–ORGANIC MATTER INTERACTIONS

Much of the organic matter in soil is complexed in some way with the clay minerals and free oxides; knowledge of these interactions is therefore of great importance to the understanding of virtually every process—physiochemical and biological—which occurs in the soil. Broadly speaking, the organic compounds may be divided into cationic, anionic and nonionic compounds of low (< 1000) to high ($> 100,000$) molecular weight. Because of the polarity of charge on clays below pH 7, organic cations are adsorbed mainly on the planar faces and anions at the edges, except that small organic anions can form links with exchangeable cations on the cleavage faces. Uncharged organic molecules interact with the surfaces through polar groups and non-specific van der Waals' forces.

Electrostatic interactions

CATIONS
Many organic cations are formed by protonation of amine groups, as in the alkyl* amines and amino acids, according to the reactions:

$$R\text{–}NH_2 + H^+\text{clay} \rightleftharpoons R\text{–}NH_3^+ \text{ clay} \qquad (7.10)$$
$$\text{(alkyl amine)}$$

and

$$\underset{\substack{| \\ COO^-}}{R\text{—}CH\text{—}NH_2} + 2H^+ \text{ clay} \rightleftharpoons \underset{\substack{| \\ COOH}}{R\text{—}CH\text{—}NH_3^+} \text{ clay} \qquad (7.11)$$

(amino acid)

Protonation is facilitated near clay surfaces by the presence of exchangeable cations which increase the

* A chain hydrocarbon group such as CH_3, C_2H_5 . . . R

dissociation of their associated water molecules and hence the acidity of the surface water. On dry clay surfaces the polarizing effect of the metallic cations is concentrated on fewer water molecules so that the degree of dissociation is increased and the protonation of adsorbed organic molecules correspondingly enhanced. The displacement of the metallic cations by organic cations such as the alkyl amines reduces the affinity of the clay surface for water, making the soil more water-repellant and diminishing the swelling propensity of the clay.

Organic cation adsorption takes place at the internal and external surfaces of 2:1 expanding-lattice clays. The adsorption affinity increases with the chain length of the compound due to the van der Waals' forces between the adsorbed molecules, so that adsorption in excess of the *CEC* may be achieved. Proteins adsorbed in inter-lamellar regions are less susceptible to microbial attack; conversely, extracellular enzymes, which are functional proteins, often show reduced biochemical activity in the adsorbed state.

ANIONS

Organic compounds are also adsorbed due to the attraction of ($-COO^-$) groups to positive edge faces, or by the displacement of OH^- and H_2O ligands of polyvalent exchangeable cations (see *cation bridge* of Figure 3.12). Phenolic-type compounds like tannins which are active constituents of the litter of many coniferous trees form co-ordination complexes with Fe and Al, and in small concentrations are very effective *deflocculants* of clays. The compound must contain at least 3 phenolic groups, of which two co-ordinate the metal atom and the third dissociates to confer a negative charge on the whole complex. Conversely, long chain polyanions are particularly effective *flocculants* of clays, since the clay particle edges become attached to the anionic groups of the molecule in the manner of a 'string of beads' (Figure 7.18). Such compounds have therefore been applied successfully, if expensively, as *soil conditioners* to improve soil structure (see Table 4.2).

Specific and non-specific van der Waals' forces and entropy stabilization

Low molecular weight alcohols, sugars and oligosaccharides are not adsorbed in competition with water unless present in large concentrations. Highly polar compounds can interact with the surface by forming

(a)

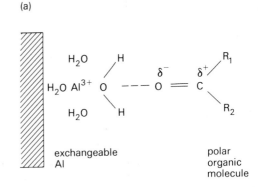

A "water bridge"

Figure 7.19. Possible interactions between organic compounds and clay surfaces
(a) A 'water bridge'

ion-dipoles or co-ordination complexes with the exchangeable cations, the link forming through the water of hydration, which is called a 'water bridge' (Figure 7.19a), or by direct hydrogen-bonding to another organic compound that is already adsorbed (Figure 7.19b).

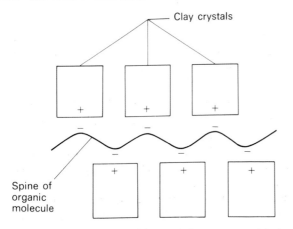

Figure 7.18. 'String of beads' arrangement of organic polyanion and kaolinite crystals.

(b)

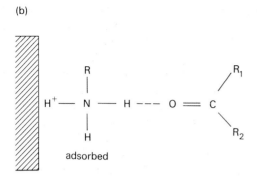

Hydrogen-bonding between
two organic compounds

Figure 7.19.
(b) Hydrogen-bonding between two organic compounds.

High molecular weight alcohols (e.g. polyvinyl alcohol) and polysaccharides are strongly adsorbed at clay surfaces, even in competition with water. The attraction occurs partly through hydrogen-bonding to O and OH groups in the surface and partly through non-specific van der Waals' forces. But the greatest contribution to the adsorption energy probably comes from the large increase in entropy when the organic macromolecule displaces many water molecules from the surface—the entropy increase stabilizes the molecule in the adsorbed state.

7.6 SUMMARY

Retention of cations and anions, soil acidity, water adsorption, and the formation and stabilization of structure are but a few of the many soil properties that are determined by reactions occurring at mineral and organic surfaces.

Apart from their large specific areas, these surfaces are important because of the charge they bear—either a *permanent negative charge* due to isomorphous substitutions in the crystal lattice, or a *pH-dependent charge* due to reversible dissociation of protons from OH and OH_2^+ groups (clay mineral edges and oxides) or COOH and ◎—OH groups of organic matter. The preferential attraction of cations and repulsion of anions at planar clay surfaces give rise to an *electrical double layer*, comprising Stern layer cations tightly adsorbed at the surface plus loosely bound cations in the diffuse or Gouy layer extending into the solution.

The repulsive force developed as diffuse layers of adjacent clay crystals overlap is manifest as a *swelling pressure* which can disrupt the microstructure of clay domains. Attractive forces between clay plates include non-specific van der Waals' forces and electrostatic forces emanating from shared divalent cations or sesquioxide films. When such forces exceed the repulsive force, *face-to-face* flocculation of the clay ensues; alternatively, the crystals may flocculate *edge-to-face*, as in the case of kaolinite at pH < 7.

The sum of the cation equivalents in the double layer, at a reference pH of 7, comprises the *cation exchange capacity* (*CEC*). The difference between the *CEC* and $\sum(Ca, Mg, K, Na)$ defines the *exchange acidity*, and the ratio

$$\frac{\sum(Ca, Mg, K, Na)}{CEC} \times 100$$

defines the *per cent base saturation*. Both Al^{3+} and H^+ contribute to exchange acidity. Hydrolysis of Al^{3+} and polymerization of the hydroxy-Al products accounts for much of the *pH buffering capacity* of a soil over the pH range 4.0 to 7.5.

Cation exchange between the double layer and the soil solution may be described by the *Gapon equation*, e.g.

$$\frac{K\text{-exchanger}}{Ca\text{-exchanger}} = k_{K-Ca} \frac{H^+}{\sqrt{Ca^{2+}}}$$

Changes in solution activities that occur on leaching soils, or in those influenced by saline water, may be explained in terms of the *ratio law*, which predicts that when cations in solution are in equilibrium with a larger number of exchangeable cations, the *activity ratio* $a^+/(a^{n+})^{\frac{1}{n}}$ is a constant over a limited range of salt concentration.

The anions $H_2PO_4^-$, HCO_3^-, $SO_4^=$ and Cl^- are adsorbed at pH-dependent sites when these are positively charged, and phosphate has a very high affinity for sesquioxide surfaces. Organic anions compete at such sites, and overall are important in linking together clay domains, sesquioxides and larger particles to form stable micro-aggregates.

REFERENCES

SCHOFIELD R.K. (1947a) A ratio law governing the equilibrium of cations in the soil solution. *Proceedings of the 11th International Congress of Pure and Applied Chemistry (London)* **3**, 257–261.

SCHOFIELD R.K. (1947b) Calculation of surface areas from measurements of negative adsorption. *Nature, London* **160**, 408.

FURTHER READING

GREENLAND D.J. (1965) Interaction between clays and organic compounds in soils. I. Mechanisms of interaction between clays and defined organic compounds. *Soils and Fertilizers* **28**, 415–425.

GREENLAND D.J. (1965) Interaction between clays and organic compounds in soils. II. Adsorption of soil organic compounds and their effect on soil properties. *Soil and Fertilizers* **28**, 521–532.

MORTLAND M.M. (1970) Clay-organic complexes and interactions. *Advances in Agronomy* **22**, 75–117.

QUIRK J.P. (1968) Particle interaction and soil swelling. *Israel Journal of Chemistry* **6**, 213–234.

THOMAS G.W. (1977) Historical developments in soil chemistry: ion exchange. *Soil Science Society of America Journal* **41**, 230–238.

VAN OLPHEN H. (1977) *An Introduction to Clay Colloid Chemistry*, 2nd Ed. Wiley Interscience, New York.

Chapter 8
Soil Aeration

8.1 SOIL RESPIRATION

Respiratory quotient and respiration rate

The importance of soil organisms in promoting the turnover of carbon has been outlined in Chapter 3. *Aerobic respiration* involves the breakdown or dissimilation of complex C molecules, oxygen being consumed and CO_2, water and energy for cellular growth

being released. Under such conditions, the *respiratory quotient*, defined as

$$RQ = \frac{\text{volume of } CO_2 \text{ released}}{\text{volume of } O_2 \text{ consumed}} \quad (8.1)$$

is equal to 1. When respiration is anaerobic, however, the RQ rises to infinity since O_2 is no longer consumed but CO_2 continues to be evolved. The respiratory

Figure 8.1. Diagram of a Rothamsted soil respirometer (after Currie, 1975).

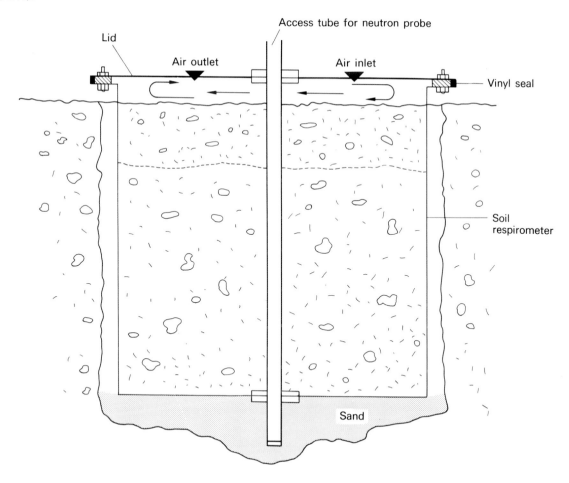

activity of the soil organisms or *respiration rate R* is measured as the volume of O_2 consumed (or CO_2 released) per unit soil volume per unit time.

Soil respiration is augmented by the respiration of living plant roots, and further, the respiration of micro-organisms is greatly stimulated by the abundance of carbonaceous material (mucilage, sloughed-off cells and exudates) in the soil immediately around the root. This zone which is influenced by the root is called the *rhizosphere*. Respiration rates and *RQ* values representative of a field soil in the south of England are given in Table 8.1.

Table 8.1. Soil respiration as influenced by crop growth and temperature.

	Respiration rate, *R*			
	January		July	
	Fallow	Cropped	Fallow	Cropped
		(1 m^{-2} day^{-2})		
O_2 demand	0.5	1.4	8.1	16.6
CO_2 release	0.6	1.5	8.0	17.4
RQ	1.20	1.07	0.99	1.05

(after Currie, 1970)

Measurement of respiration rates

An instrument designed to measure respiration rates is called a *respirometer*, and one designed specifically for field measurements is illustrated in Figure 8.1. The soil moisture content is regulated by controlled watering and drainage; temperature is recorded and the circulating air is monitored for O_2, the deficit being made good by electrolytic generators. Carbon dioxide is absorbed in vessels containing soda lime and measured.

Such measurements show that the respiration rate depends on
(a) soil conditions such as organic matter content and moisture;
(b) cultivation and cropping practices;
(c) environmental factors, principally temperature.

The effect of temperature on respiration rate is expressed by the equation:

$$R = R_0\, Q^{T/10} \tag{8.2}$$

where *R* and R_0 are the respiration rates at temperature *T* and 0°C respectively, and *Q* is the magnitude of the increase in *R* for a 10° rise in temperature, called the *Q-10 factor*, which lies between 2 and 3. Temperature change is the cause of large seasonal fluctuations in soil respiration rate in temperate climates, as illustrated in Figure 8.2.

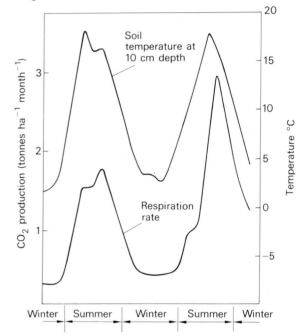

Figure 8.2. Seasonal trends in soil temperature and respiration rate in southern England (after Currie, 1975).

Cycles of respiratory activity

In southern England the seasonal maxima in *R* value lag 1–2 months behind those of soil temperature. Superimposed on this seasonal pattern is the effect of weather, for within any month of the year the daily maximum in *R* may be up to twice the daily minimum, a fluctuation strongly correlated with soil surface or air temperatures. Finally, there are minor variations in *R* caused by the diurnal rise and fall of soil temperature.

These trends are modified by other factors—for example, at Rothamsted in southern England, the respiratory peak is consistently lower during a dry summer than a wet one, and the rate tends to be higher in spring than in autumn for the same average soil temperatures. The latter effect is attributable to two factors, primarily a greater supply of organic residues at the end of the quiescent winter period than after a summer of active decomposition, together with spring cultivations which break up large aggregates and expose fresh organic matter to attack by micro-organisms.

8.2 MECHANISMS OF GASEOUS EXCHANGE

The composition of the soil air, as described in section 4.5, is buffered against change by the much larger volume of the Earth's atmosphere. Buffering is achieved by the dynamic exchange of gases as O_2 travels from the atmosphere to the soil and CO_2 moves in the reverse direction, a process called *soil aeration*. Three transport processes are involved in soil aeration:
(a) Dissolved O_2 is carried into the soil by percolating rainwater; the contribution is small owing to the low solubility of O_2 in water (0.028 ml ml^{-1} at 25°C and 1 bar pressure).
(b) Mass flow of gases due to pressure changes of 1–2 mbars created by wind turbulence over the surface.
(c) Diffusion of gas molecules through the soil pore space.

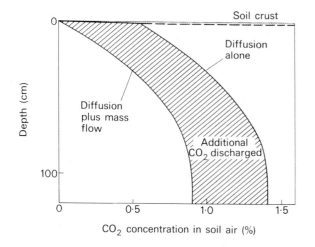

Figure 8.3. Concentration *vs* depth profiles for CO_2 in the soil air under a surface crust (after Currie, 1970).

Gaseous diffusion is far more important than mass flow, even when the soil surface is sealed by structure breakdown. As Figure 8.3 illustrates, the CO_2 concentration in the air space of a 'capped' soil is reduced to less than 1.5 per cent at 1 m depth by diffusion alone, with a further improvement of only 0.5 per cent when mass flow across the thin surface crust occurs.

Gas exchange by diffusion

STEADY-STATE CONDITIONS
The steady-state rate of diffusion, or *diffusive flux F*, is given by the equation:

$$F = -D \ grad \ C \qquad (8.3)$$

where F is the quantity of gas diffusing across a unit area perpendicular to the direction of movement in unit time; *grad C* is the change in gas concentration C per unit length in the direction of diffusion, and D is a proportionality constant called the *diffusion coefficient* (compare with equation 6.7 for water flow). Values of D for CO_2 and O_2 in water and air are given in Table 8.2, from which it is seen that for the same *grad C* the rate of gas diffusion is 10,000 times faster through air-filled pores than water-filled pores. Further, since D_{O_2} is slightly larger than D_{CO_2}, when the soil RQ is 1, O_2 diffuses into the soil faster than the CO_2 that is produced can diffuse out. A pressure difference builds up and mass flow of N_2 occurs to eliminate it. Nevertheless, under steady-state aerobic conditions it is sufficiently accurate to equate the diffusive flux of O_2 into the soil with that of CO_2 out.

More significantly, the D values for O_2 and CO_2 through the air-filled pores are always less than the D values in the outside air because of the restricted volume of the pores and the tortuosity of the diffusion pathway.

Table 8.2. Diffusion coefficients of O_2 and CO_2 in air and water.

	Oxygen	Carbon dioxide
Air	2.3×10^{-1}	1.8×10^{-1}
Water	2.6×10^{-5}	2.0×10^{-5}

These effects are expressed in the equation:

$$D = \alpha \, \epsilon \, D_o \qquad (8.4)$$

where D and D_o are the diffusion coefficients of the gas in the soil pores, and in the outside air respectively, ϵ is the air-filled porosity (section 4.5) and α is an *impedance factor* which decreases as the pathway becomes more tortuous.

The value of α varies in a complex way with soil structure and water content as illustrated in Figure 8.4. The change in α in dry sand (Figure 8.4a) is limited by

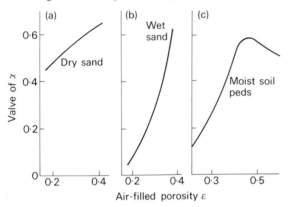

Figure 8.4. The relation between impedance factor α and air-filled porosity ϵ for dry and wet sand and soil peds (after Currie, 1970).

the extent to which the pore-size distribution within the matrix can be altered by changing the packing density and by the addition of finer particles; but when the sand is wet (Figure 8.4b), water films fill the narrowest necks between pores so that the shape as well as the size distribution of the pores is altered. When water is added to dry soil peds (Figure 8.4c), α initially increases as the *intraped* pores fill and the tortuosity and roughness of the gas diffusion path actually decreases. However, once the larger *interped* pores begin to fill, α declines rapidly as it does in the wet sand, approaching a minimum value at $\epsilon \simeq 0.1$, a value corresponding to the average proportion of dead-end pores in the soil. We may conclude that in structured soils, the intraped pores make little contribution to gas diffusion and the *major* pathway is through the interped or macropores, most of which are drained at water contents less than the field capacity. The value of α can be as high as 0.6 in dry, well struc-

tured soils and as low as 0.16 in over-worked arable soils.

NON-STEADY STATE CONDITIONS

Because the concentrations of O_2 and CO_2 in the pore space of field soils are changing in space and time, the exchange of gases occurs under non-steady state conditions. The movement of O_2 and CO_2 must then be described by combining the diffusive flux equation (equation 8.3) with the continuity equation, which gives

$$\frac{dC}{dt}\epsilon = -\frac{d(F)}{dz} \pm R$$

i.e.

$$\frac{dC}{dt}\epsilon = D_z \frac{d^2C}{dz^2} \pm R \qquad (8.5)$$

where D_z is the *effective* D value in the z direction (vertical) and R is the rate of CO_2 production ($+R$), or O_2 consumption ($-R$), usually expressed as ml gas cm^{-3} soil s^{-1}.

Assuming that D_z is independent of depth and time, and R is not greatly affected by O_2 concentration, a solution to equation 8.5 is obtained, putting $dC/dt = 0$, which relates the change in O_2 concentration (ΔC) to the depth over which change occurs; that is,

$$\Delta C \simeq \frac{Rz^2}{2D_z} \qquad (8.6)$$

Substitution of appropriate values for R, z and D_z in this equation gives changes in O_2 concentration generally consistent with field measurements: two contrasting situations serve to illustrate the point.

(a) For a dry sandy soil, putting $\epsilon = 0.4$ and $\alpha = 0.6$ in equation 8.4 gives $D_z(O_2) = 5.5 \times 10^{-2}\ cm^2\ s^{-1}$, and taking $R \simeq 2 \times 10^{-7}$ ml $cm^{-3}\ s^{-1}$ (Greenwood, 1975), gives

$\Delta C = 0.004$ ml ml^{-1} gas phase at a depth of 50 cm

This corresponds to a negligible fall in O_2 partial pressure of 0.004 bars, by Dalton's law of partial pressures (section 4.5).

(b) On the other hand, in a soil where poor structure and consequent waterlogging impede diffusion in the gas phase, D_z may fall to $1.5 \times 10^{-3}\ cm^2\ s^{-1}$, giving

$\Delta C = 0.166$ ml ml^{-1} gas phase at 50 cm,

and an O_2 partial pressure of $0.21 - 0.17 = 0.04$ bars.

These calculated results agree well with measured O_2 concentrations for freely and poorly drained soils, shown in Figure 8.5. The restriction on gas diffusion imposed by the predominance of water-filled pores in a poorly drained soil is discussed below.

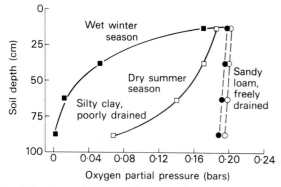

Figure 8.5. Changes in O_2 partial pressure in contrasting soils with depth and season of the year.

GAS DIFFUSION THROUGH WATER FILMS
The enzyme cytochrome oxidase, which catalyses the reduction of O_2 in plants and micro-organisms, has such a high affinity for O_2 that the rate of aerobic respiration only falls appreciably at O_2 partial pressures <0.01 bars. Although partial pressures above this value may be maintained in the few air spaces remaining in a water-logged soil, the respiring organisms live in the water-filled pores *within* peds so that O_2 must diffuse through water, a slow process due to the low solubility of oxygen especially at low partial pressures. In many soils, therefore, anaerobic pockets develop at the centre of peds larger than a certain size if the peds remain wet for any length of time.

To illustrate this effect, equation 8.5 can be solved, assuming the peds to be spherical and that R and D do not change rapidly with time, to give the equation:

$$a^2 = \frac{6\Delta CD}{R} \qquad (8.7)$$

where a is the ped radius, ΔC is the difference in O_2 concentration between the ped surface and its centre, D is the effective diffusion coefficient of O_2 *through the water phase* and R the respiration rate inside the ped.

Since the gas-filled pores of the soil are usually continuous at $\epsilon > 0.1$, a partial pressure of O_2 of 0.21 bars at the surface of the ped may be assumed, giving the concentration of O_2 in the water there as

$$0.21 \times 0.028 = 0.0059 \text{ ml } O_2 \text{ ml}^{-1} \text{ water at } 25°C$$

Putting $D = 1 \times 10^{-5}$ cm^2 s^{-1} (Greenwood, 1975), equation 8.7 is solved for different values of R to give the maximum radius of a ped, the centre of which will just be anaerobic. The results are shown in Figure 8.6.

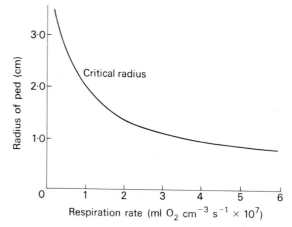

Figure 8.6. Critical radius for the induction of anaerobic conditions at ped centres (after Greenwood, 1975).

It follows that provided the soil has an air-filled porosity > 0.1, anaerobic conditions are unlikely to occur unless there are sites of active respiration separated from the gas phase by more than 1 cm of water-saturated soil. This is probably a common occurrence in heavy clay soils in Britain during the winter and spring months. Taking the same condition of a wet ped of radius a, just less than the critical radius for the onset of anaerobic conditions, one may calculate the maximum accumulation of CO_2 within the ped by equating the influx of O_2 with the efflux of CO_2. Using equation 8.3 we have:

$$D_{O_2}\frac{\Delta C'}{a} = D_{CO_2}\frac{\Delta C''}{a} \qquad (8.8)$$

i.e.

$$\Delta C'' = \frac{D_{O_2}}{D_{CO_2}}\Delta C' \qquad (8.9)$$

where $\Delta C'$ and $\Delta C''$ are the concentration differences in the water phase for O_2 and CO_2 respectively. Substituting in equation 8.9 for D_{O_2} and D_{CO_2} from Table 8.2, and putting $\Delta C' = 0.0059$, gives

$$\Delta C'' = 0.0077$$

Since the solubility of CO_2 at 25°C and 1 bar pressure equals 0.759, it follows that the CO_2 partial pressure at the ped centre is $c.\ 0.01$ bars. An increase in partial pressure of this magnitude has a negligible effect on plant and microbial metabolism. Nevertheless, if the O_2-free zones in the soil are extensive, as in a waterlogged soil, the CO_2 partial pressure may rise to 10 times this value.

8.3 EFFECTS OF POOR SOIL AERATION ON ROOT AND MICROBIAL ACTIVITY

Plant root activity

The picture emerging from the previous discussion is one of a mosaic of aerobic and anoxic zones in the soil, in the case of heavy textured soils during wet periods of the year, as illustrated in Figure 8.7. The anoxic zones

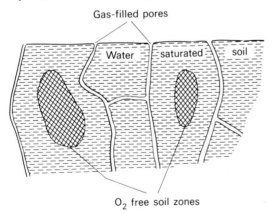

Figure 8.7. Distribution of water and oxygenated zones in a structured soil (after Greenwood, 1969).

exert no ill effect (other than through the reduction of NO_3^-, discussed later) as most of the roots grow through the larger pores ($> 100\ \mu m$ diameter) where O_2 diffusion is rapid. Oxygen can also diffuse for limited distances through plant tissues from regions of high to low O_2

supply, an attribute best developed in aquatic species and lowland rice (*Oryza sativa*), because of larger intercellular air spaces in which the value of D_{O_2} approaches 1×10^{-1} cm^2 s^{-1}.

However, when heavy rain coincides with temperatures high enough to maintain rapid respiration rates (typically in spring), the *whole* root zone can become anaerobic, a condition which even if sustained for only a day, has serious effects on the metabolism and growth of sensitive plants. Species intolerant of waterlogging experience an acceleration in the rate of glycolysis and the accumulation of endogenously produced ethanol; cell membranes become leaky, and ion and water uptake is impaired. The characteristic symptoms are wilting, yellowing of leaves and the development of adventitious roots at the base of the stem. Stimulation of the production of endogenous ethylene (C_2H_4) may also cause growth abnormalities such as leaf epinasty, the downward curvature of a leaf axis.

Soil microbial activity

Soil aeration determines the balance between aerobic and anaerobic metabolism which profoundly affects the energy available for microbial growth and the type of end-products. *Aerobic organisms* are generally beneficial to soil and plants—the *heterotrophs* produce much biomass, water and CO_2, and unlock essential elements such as N, P and S from organic combination; while specialized groups of *autotrophs*, such as the nitrifying bacteria and sulphur-oxidizing bacteria produce NO_3^- and $SO_4^=$. Some of the rhizosphere-dwelling species of bacteria produce growth regulators, including gibberellic acid and indolacetic acid, which may stimulate plant growth. By contrast, *anaerobic organisms* produce less cell growth but appreciable amounts of simple organic compounds, notably low molecular weight organic acids, CO_2, methane (CH_4) and volatile amines. They also initiate chemical reactions that are often undesirable, such as the reduction of NO_3^-, MnO_2 and $SO_4^=$ (section 8.4).

AEROBIC PROCESSES: NITRIFICATION

Nitrification entails the biological oxidation of inorganic N forms to nitrate (NO_3^-). Nitrification is discussed here because it is an essential prior step to the loss of

N by NO_3^- reduction and the evolution of N_2O and N_2—a process called *denitrification*.

The principal nitrifying organisms are chemoautotrophic bacteria of the genera *Nitrosomonas* and *Nitrobacter*. These organisms are unique in deriving energy for growth solely from the oxidation of NH_4^+ produced by saprophytic decay organisms, or accruing to the soil from fertilizers and rain. The oxidation occurs in two steps:

(a) $NH_4^+ + \frac{3}{2}O_2 \rightleftharpoons NO_2^- + 2H^+ + H_2O + \text{energy}$ (8.10)

carried out by the *Nitrosomonas* group, and

(b) $NO_2^- + \frac{1}{2}O_2 \rightleftharpoons NO_3^- + \text{energy}$ (8.11)

carried out by *Nitrobacter*. These energy-releasing reactions are coupled to cellular synthesis and to the reduction of CO_2 which is the sole source of C for these organisms. As CO_2 is plentiful, the growth rate and hence bacterial numbers are limited primarily by the supply of oxidizable substrates: for example, the demand for NH_4^+ created by the more numerous soil heterotrophs restricts the rate of NH_4^+ oxidation by the slow growing nitrifiers, and the absence of detectable NO_2^- in well aerated soils attests to the limitation that NO_2^- supply imposes on *Nitrobacter* numbers.

Growth of the organisms and rate of nitrification is also influenced by temperature, moisture, pH and especially O_2 supply. The *temperature* optimum lies 30 and 35°C, but some nitrification proceeds at temperatures down to 0°C. The optimum *moisture content* is approximately 60 per cent of field capacity. *Nitrosomonas* is more susceptible to dry conditions than *Nitrobacter*, but sufficient bacteria survive short periods of desiccation for increased nitrification to follow the flush of organic decomposition contingent upon the re-wetting of an air-dry soil. An alkaline *soil pH* is most favourable, *Nitrosomonas* attaining peak activity between pH 7 and 9 and *Nitrobacter* having a somewhat lower optimum (section 12.2).

ANAEROBIC PROCESSES: FERMENTATION AND REDUCTION PRODUCTS

The pathway of cellular oxidation of carbohydrates is the same in plants and micro-organisms, in the presence and absence of O_2, as far as the key intermediate,

Figure 8.8. Products of the fermentation of carbohydrates (after Yoshida, 1975).

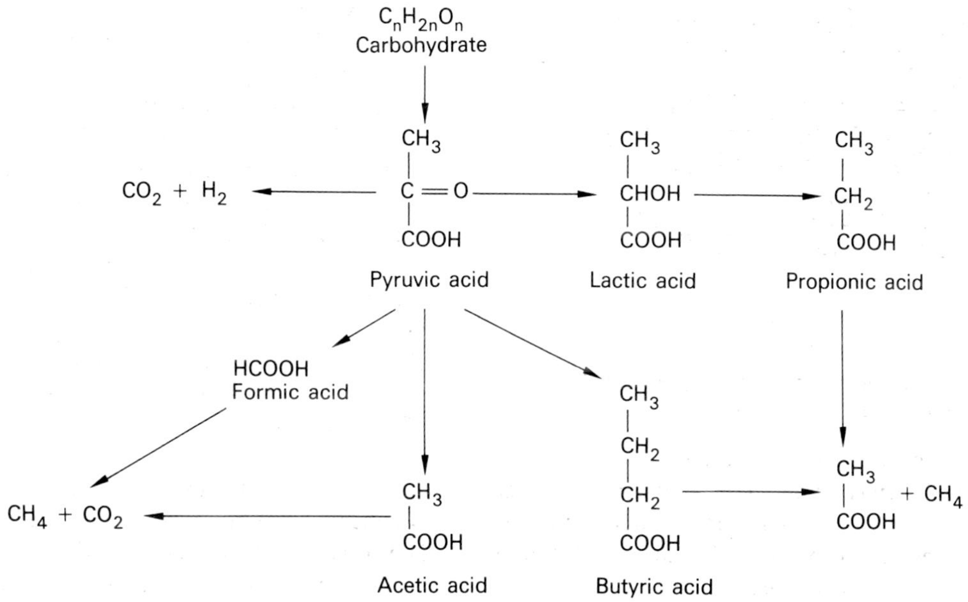

pyruvic acid ($CH_3COCOOH$). This is called *glycolysis*. Normally, reduced coenzymes in the cell are reoxidized as electrons are passed along a chain of respiratory enzymes to the terminal acceptor, oxygen. In the absence of O_2, however, there are two possible alternatives:

(a) Other organic compounds serve as final electron acceptors and are reduced in turn—the process of *fermentation*;

(b) inorganic compounds such as NO_3^-, $Fe(OH)_3$, MnO_2 and $SO_4^=$ are reduced (section 8.4).

As indicated in Figure 8.8, several low M.W. or volatile fatty acids (VFA) are formed as intermediates in carbohydrate fermentation: in the soil solution around fermenting crop residues they can attain concentrations of 1–10 mM which are harmful to plants, especially young seedlings. The VFAs are finally dissimilated to the simple gases CO_2, CH_4 and H_2 by obligate anaerobes. Because anaerobic respiration releases less metabolic energy than aerobic respiration, microbial growth is slower and a greater proportion of the N released on protein decomposition appears free in the soil as NH_4^+; putrefaction products such as amines, thiols and mercaptans may also accumulate.

ETHYLENE PRODUCTION

One soil organism found to produce the hydrocarbon gas ethylene (C_2H_4) is the early-colonizing sugar fungus *Mucor*, primarily *Mucor hiemalis*, which only grows aerobically; however, it appears that transient or localized anaerobic conditions are necessary to provide suitable substrates for C_2H_4 production. Very low concentrations of ethylene (~ 0.1 ppm) stimulate root elongation, but concentrations of 1 ppm and greater, which may occur in wet, poorly drained soil or beneath the smeared plough layer in a clay, seriously inhibit root growth (Figure 8.9). Some micro-organisms also metabolize C_2H_4, there being coryneform bacteria that consume C_2H_4 as their sole source of C. The ethylene concentration therefore reflects the balance between microbial production, and diffusive losses and decomposition, which explains why very little C_2H_4 accumulates in well aerated soils where it is destroyed or lost as rapidly as it is formed.

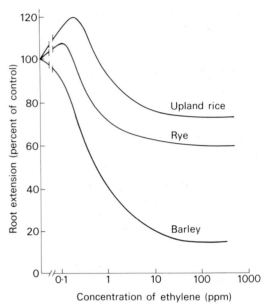

Figure 8.9. The effect of ethylene on root growth (after Cannell, ARC Letcombe Laboratory Rep., 1975).

Indices of soil aeration

From the foregoing we can see that the aeration status of a soil may be assessed in one of several ways:

(1) Measurement of the rate of O_2 diffusion through the soil. A popular method relies on the principle that reduction of O_2 at the surface of a platinum (Pt) electrode inserted into the soil, to which a fixed voltage is applied, causes a current flow proportional to the rate of oxygen diffusion to the electrode. The problems encountered in applying this method have been reviewed by McIntyre (1970). Growth restriction due to lack of O_2 probably occurs at O_2 partial pressures <0.05 bars but higher values do not necessarily mean that anaerobic zones do not exist.

(2) The detection of fermentation and putrefaction products by their odour (especially the volatile amines and mercaptans) and where possible by quantitative analysis (the VFAs, CH_4 and C_2H_4).

(3) By noting the associated chemical changes—for example, the solution and reprecipitation of Mn as small black specks of MnO_2, or the appearance of a bluish-grey colour of reduced Fe compounds.

Once the O_2 supply in the soil has been depleted, and anaerobic processes are initiated, the *redox potential* is a useful measure of the intensity of the ensuing reducing conditions.

8.4 OXIDATION-REDUCTION REACTIONS IN SOIL

Redox potential

The final step in aerobic respiration is the reaction:

$$O_2(gas) + 4H^+(soln) + 4e \rightleftharpoons 2H_2O(liquid) \qquad (8.12)$$

which is an example of an oxidation-reduction or *redox reaction*. The general form of this reaction is

$$Ox + mH^+ + ne \rightleftharpoons Red \qquad (8.13)$$

for which [Ox] and [Red] are the activities of the oxidized and reduced species respectively. By combining the equation for the free energy change in a reversible reaction

$$\Delta G = \Delta G^o + RT\ln \frac{[Products]}{[Reactants]} \qquad (8.14)$$

with the equation for electrical work done for a given free energy change (ΔG)

Figure 8.10. Threshold redox potentials at which the oxidized species shown become unstable (after Patrick and Mahapatra, 1968).

$$\Delta G = -nFE \qquad (8.15)$$

where n = number of electrons transferred through an electrical potential difference E and F is Faraday's constant, we may write

$$E = E^o - \frac{RT}{nF}\ln\left\{\frac{[Red]}{[Ox][H^+]^m}\right\} \qquad (8.16)$$

i.e.

$$E_h = E^o + 2.3\frac{RT}{nF}\left\{\log_{10}\frac{[(Ox)]}{[Red]} - mpH\right\} \qquad (8.17)$$

In equation (8.17), E_h measures the *equilibrium redox potential* for known activities of H^+ ions and the oxidized and reduced species, while E^o is the potential of a chosen redox system ($H^+ + e \rightleftharpoons \frac{1}{2}H_2$) under standard conditions (unit activities of the participants at 25°C and a pressure of 1 atmosphere). Clearly, oxidizing systems will have relatively positive E_h values and reducing systems more negative E_h values. It also follows that for a given ratio of [Ox]/[Red], the lower the pH the higher the E_h value at which reduction occurs.

In practice, redox potentials are measured as the electrical potential difference between a Pt electrode

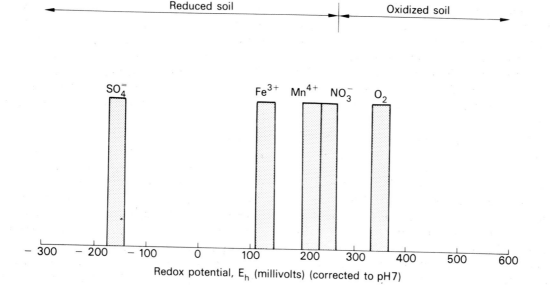

and a reference electrode (calomel) which are placed in the solution in equilibrium with the soil. Reproducible results are only obtained in O_2-free systems, wherein characteristic redox reactions are associated with relatively narrow ranges of E_h. Each reaction buffers or *poises* the system in the particular E_h range until that oxidant is exhausted, and the E_h drops to a lower value as reducing power intensifies.

Sequential reductions in a waterlogged soil

When a soil becomes submerged or waterlogged, the O_2 present is consumed within 1 to 2 days, and if anaerobic conditions persist eventually NO_3^-, Mn^{4+}, Fe^{3+} and $SO_4^=$, in that order, will be reduced. The approximate E_h values at which the soil is successively poised by each of these redox systems are shown in Figure 8.10. The sequence of reactions that occurs is:

$$\frac{1}{5}NO_3^-(soln) + \frac{6}{5}H^+(soln) + e \rightleftharpoons \frac{1}{10}N_2(gas) + \frac{3}{5}H_2O$$
$$\text{(liquid)} \qquad (8.18)$$

$$\frac{1}{2}MnO_2(solid) + 2H^+(soln) + e \rightleftharpoons \frac{1}{2}Mn^{2+}(soln) + H_2O$$
$$\text{(liquid)} \qquad (8.19)$$

$$Fe(OH)_3(solid) + 3H^+(soln) + e \rightleftharpoons Fe^{2+}(soln + 3H_2O$$
$$\text{(liquid)} \qquad (8.20)$$

$$\frac{1}{8}SO_4^=(soln) + \frac{5}{4}H^+(soln) + e \rightleftharpoons \frac{1}{8}H_2S(gas) + \frac{1}{2}H_2O$$
$$\text{(liquid)} \qquad (8.21)$$

DENITRIFICATION

The first reaction (8.18) summarizes the process of *denitrification* which is carried out by facultative anaerobic bacteria in the soil once the partial pressure of O_2 has fallen to a very low level (at pH 7 and a NO_3^- concentration of 10^{-3} M, reduction to N_2 gas requires the O_2 partial pressure to fall to 10^{-7} bars). The complete pathway for NO_3^- reduction is as follows:

$$NO_3^- \rightleftharpoons NO_2^- \rightleftharpoons N_2O \rightleftharpoons N_2$$
nitrous
oxide

aerobic   anaerobic

When the rate of denitrification is slow (organic substrates and nitrate limiting) little N_2O can be detected in the soil, but under conditions favouring rapid reduction the ratio of N_2O/N_2 in the gas evolved rises markedly.

Denitrification losses can be serious in agricultural soils, especially those experiencing spatial changes from aerobic to anaerobic conditions, and the reversal of these conditions with time. An extreme situation can arise in lowland rice culture, as illustrated in Figure 8.11,

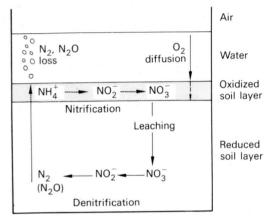

Figure 8.11. Nitrification and denitrification in the oxidized and reduced zones of a flooded soil (after Patrick and Mahapatra, 1968).

in which the NO_3^- produced aerobically diffuses or is leached into anaerobic zones where it is denitrified. Accordingly, N fertilizer in such systems is best applied in an NH_4^+ form and placed directly in the anoxic zone below the soil surface.

More intensive reducing conditions

Reactions 8.19 and 8.20 are characteristic of *gleying*, a pedogenic process discussed in section 9.2. The former is less effective in poising the soil than either NO_3^- or Fe^{3+} reduction because MnO_2 is very insoluble and relatively few micro-organisms use it as a terminal electron acceptor. Nevertheless, in very acid soils waterlogging may lead to high levels of exchangeable Mn^{2+} which cause manganese toxicity in susceptible plants. Anaerobiosis and Fe^{3+} reduction generally increase the phosphate concentration in the soil solution especially in acid soils where phosphate is immobilized in very insoluble Fe compounds. The effect is complicated, however, by the inevitable rise in pH on prolonged waterlogging ($3H^+$ ions are consumed for each electron transferred in reaction 8.20): at pH > 6 ferrous compounds become less soluble and an oxide containing Fe in the ferric and ferrous states precipitates, probably hydrated magnetite $Fe_3(OH)_8$, which presents a highly active surface for the readsorption of phosphate. Certainly all the dissolved P is readsorbed when aerobic conditions return and amorphous $Fe(OH)_3$ precipitates.

At very low E_h values (Figure 8.10), $SO_4^=$ is reduced due to the activity of obligate anaerobes of the genus *Desulphovibrio*. As the concentration of weakly dissociated H_2S builds up in the soil solution, ferrous sulphide (FeS) precipitates and slowly reverts to iron pyrites (FeS_2), a mineral characteristic of marine sedimentary rocks and of sediments currently deposited in river estuaries.

8.5 SUMMARY

During respiration by soil organisms and plant roots, complex C molecules are dissimilated to provide energy for cellular growth. When conditions are *aerobic*, one mole of CO_2 is released for each mole of O_2 consumed and the *respiratory quotient* (*RQ*) is 1. Respiration continues aerobically, at a rate determined by soil moisture, temperature and substrate availability, provided that O_2 from the atmosphere moves into the soil fast enough to satisfy the organisms' requirements, and CO_2 can move out. This dynamic exchange of gases is called *soil aeration*.

Diffusion of gases through the air-filled pore space (ϵ) is the most important mechanism of soil aeration. The rate of diffusion or diffusive flux F is given by

$$F = -D \, grad \, C$$

but under non-steady state conditions (gas concentrations C changing in space and time), the time rate of change in concentration in the soil pores is given by

$$\epsilon \frac{dC}{dt} = D_z \frac{d^2C}{dz^2} \pm R$$

where R, the soil respiration rate, is positive when measured as CO_2 release and negative when measured as O_2 uptake. D_z is the gas diffusion coefficient in the vertical (z) direction.

Sandy soils with ϵ values > 0.1 maintain O_2 partial pressures close to the atmospheric level of 0.21 bars except if completely waterlogged. However, in clay soils that remain at field capacity for many weeks (English winter conditions), sites of active microbial respiration within peds become *anaerobic* if separated from air-filled pores by more than 1 cm of water-saturated soil.

Since the D values for O_2 and CO_2 in water are 10,000 less than in air, gas exchange through water-filled pores is very slow and total anaerobiosis develops when a soil is waterlogged for more than 1 to 2 days. Facultative followed by obligate anaerobic bacteria proliferate and several undesirable chemical changes may ensue. Carbohydrates are incompletely oxidized or fermented to *volatile fatty acids* which may attain localized concentrations (1–10 mM) high enough to be toxic to seedlings. In the absence of O_2, other substances serve as electron acceptors in respiration, and reduction of NO_3^-, MnO_2, Fe^{3+} compounds and $SO_4^=$ occurs in that sequence as reducing conditions intensify. The reducing power of the soil system is measured by the *redox potential E_h*, given by the equation:

$$E_h = E^0 + \frac{2.3RT}{nF}\left\{ \log_{10} \frac{[Ox]}{[Red]} - mpH \right\}$$

REFERENCES

GREENWOOD D.J. (1975) Measurement of soil aeration, in *Soil Physical Conditions and Crop Production. MAFF Bulletin* **29**, 261–272.

MCINTYRE D.S. (1970) The platinum electrode method for soil aeration measurement. *Advances in Agronomy* **22**, 235–283.

FURTHER READING

CURRIE J.A. (1970) Movement of gases in soil respiration, in *Sorption and Transport Processes in Soils. Society of Chemical Industry Monograph* **37**, 152–169.

CURRIE J.A. (1975) Soil respiration, in *Soil Physical Conditions and Crop Production. MAFF Bulletin* **29**, 461–468.

GREENWOOD D.J. (1969) Effect of oxygen distribution in soil on plant growth, in *Root Growth*. (Ed. W.J. Whittington), Butterworths, London, 202–221.

PATRICK W.H. & MAHAPATRA I.C. (1968) Transformation and availability to rice of nitrogen and phosphorus in waterlogged soils. *Advances in Agronomy* **20**, 323–359.

PONNAMPERUMA F.N. (1972) The chemistry of submerged soils. *Advances in Agronomy* **24**, 29–96.

YOSHIDA T. (1975) Microbial metabolism of flooded soils, in *Soil Biochemistry*, Volume **3**. (Eds. E.A. Paul & A.D. McLaren), Marcel Dekker, New York, 83–122.

Chapter 9
Processes in Profile Development

9.1 THE SOIL PROFILE

Soil horizons

In Chapter 1, the course of profile development from bare rock to a mature *brown forest soil* was described in outline. Except in peaty soils, the downward penetration of the weathering front greatly exceeds the upward accumulation of litter on the surface, and the mature profile may have the vertical sequence previously illustrated in Figure 1.4, that is:

Litter layer, L
A horizon (mainly eluvial)
B horizon (mainly illuvial)
C horizon (parent material)

Predictably, this simple model does not encompass the full complexity of profile development. We must turn to a more comprehensive model with an enlarged glossary of descriptive terms—the *horizon notation*, that is intended to convey briefly the maximum information about the soil to the observer. The notation chosen is that of the Soil Survey of England and Wales (Hodgson, 1974) which is consistent with current international usage.

There are two main kinds of horizon—*organic* and *mineral*, which are distinguished on their organic matter content. An organic horizon, which usually bears a superficial litter layer L, must have either

(a) greater than 30 per cent organic matter (18 per cent C) when the mineral fraction has 50 per cent or more clay; or

(b) greater than 20 per cent organic matter (12 per cent C) when the mineral fraction has no clay.

When an organic horizon remains wet most of the time, it is called a *peaty* or O horizon; otherwise it is simply called an F or H horizon, depending on the degree of humification of the plant residues (section 3.3). Wetness favours an increasing thickness of the O horizon, which qualifies as *peat* when more than 40 cm deep and the organic matter content is > 50 per cent.

Sandy peat can have between 20 and 50 per cent organic matter but must have 50 per cent or more of sand, whereas a *loamy peat* has a similar organic content but less than 50 per cent sand.

A *mineral* horizon is low in organic matter and overlies consolidated or unconsolidated rock; it may be designated A, E, B, C or G, as indicated in Table 9.1,

Table 9.1. Soil horizon notation.

Horizon	Brief description
O—organic	
Of	Composed mainly of fibrous peat
Om	Composed mainly of semi-fibrous peat
Oh	Mainly amorphous organic matter; uncultivated
Op	Amorphous organic matter mixed by cultivation
A—at or near the surface, with well mixed humified organic matter (formerly designated A_1)	
Ah	Uncultivated with $>1\%$ organic matter
Ap	Mixed by cultivation
Ag	Partially gleyed due to intermittent waterlogging; rusty mottles along root channels
E—eluvial, of low organic content, having lost material to lower horizons (formerly designated A_2)	
Ea	Lacking mottles, bleached particles due to the removal of organic and sesquioxidic coatings
Eb	Brownish colour due to uniformly disseminated iron oxides; often overlies a Bt horizon
Eg	See Ag; often overlies a Bt horizon, and sometimes a thin iron-pan
B—subsurface, showing characteristic structure, texture and colour due to illuviation of material from above and the weathering of the parent material	
Bf	Thin iron-pan or placon; often enriched with organic matter and aluminium
Bg	Partially gleyed; blocky or prismatic structure; variable black MnO_2 mottles
Bh	Translocated organic matter with some Fe and Al

Table 9.1. *continued*

Bs	Enriched with sesquioxides, usually by illuviation; orange to red colour
Bt	Accumulation of translocated clay as shown by clay coatings on ped faces
Bw	Shows alterations to parent material by leaching, weathering and structural reorganization

G—gleyed, bluey-grey to green colours predominate throughout a structureless soil matrix

C—parent material

Ck	Contains $>1\%$ secondary $CaCO_3$ as concretions or coatings (also Ak and Bk)
Cy	Contains secondary $CaSO_4$ as gypsum crystals
Cm	Continuously cemented, other than by a thin iron-pan
Cr	Weakly consolidated but dense enough to prevent root ingress except along cracks
Cu	Unconsolidated and lacking evidence of gleying, cementation etc.
Cx	With fragipan properties, as in Pleistocene deposits; dense but uncemented; firm when dry but brittle when moist

(after SSEW, 1974)

depending on its position in the profile, its mode of formation, colour,* structure, degree of weathering and other properties which can be recognized in the field. Three kinds of complex horizon are also recognized: a horizon with features of two horizons, such as a strongly gleyed C horizon, is designated C/G whereas a relatively homogeneous horizon lying between two other horizons, say A and B, is designated AB. Less commonly, a transitional horizon may comprise discrete parts (more than 10 per cent by volume) of two contiguous horizons, when it is called A & B, or E & B for example. Specific attributes of each major horizon are indicated by lower-case suffixes attached to the horizon symbol (see Table 9.1).

Soil layers

Unless the record of rock lithology or stratigraphy suggests the contrary, it is assumed that the parent material originally extended unchanged to the soil

* Standard colour descriptions are obtained from Munsell colour charts.

surface. As indicated in section 5.2, this assumption is often invalid for soils on very old land surfaces subjected to many cycles of weathering, erosion and deposition which have resulted in the layering of parent materials. An example of layering is the *K cycle* of soil formation on weathered and resorted materials in southeast Australia (Figure 9.1).

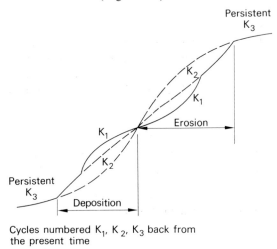

Cycles numbered K_1, K_2, K_3 back from the present time

Figure 9.1. K cycles of soil formation on a hill slope (after Butler, 1959).

Layers of different lithology are identified by numbers prefixed to the horizon symbols, the uppermost layer being 1 (prefix usually omitted) and those below labelled sequentially 2, 3 etc. It is also possible for an old soil to be buried by material of similar lithology, in which case the buried soil horizons are prefixed by the letter *b*. Buried horizons are found in very old cleared areas where man has shifted top soil for farming or construction purposes; similarly, the course of soil formation near old settlements has frequently been altered by man's deliberate deposition of organic residues (seaweed) and waste. Soils of this kind are called man-made or *plaggen soils*.

9.2 THE PRINCIPAL PROCESSES

In this section, the more important processes operative in profile development are discussed. The intensity of their effect is governed by the interaction of parent rock, climate, organisms, relief and time, subtle changes in which lead to gradations in profile development comparable to the *soil maturity sequence* which follows from the gradual loss of solutes on progressive weathering (sections 9.3–9.6).

Mineral dissolution

HYDROLYSIS

Rock minerals become unstable when exposed to a changed environment—water penetrates between the crystals and dissolves the more soluble elements. With complex minerals like the silicates, solution is aided by surface *hydrolysis*—water molecules dissociate into H^+ and OH^- and the small, mobile H^+ ions (actually H_3O^+) penetrate the crystal lattice, creating a charge imbalance which causes cations such as Ca^{2+}, Mg^{2+}, K^+ and Na^+ to diffuse out. The potash feldspar

orthoclase hydrolyses to produce a weak acid (silicic) a strong base (KOH) and leaves a residue of the clay mineral *illite*:

$$3\,K\,Al^{IV}Si_3O_8 + 14H_2O \rightleftharpoons K(Al\,Si_3)\,^{IV}Al_2^{VI}O_{10}(OH)_2$$
$$+ 6\,Si(OH)_4 + 2\,KOH \qquad (9.1)$$

Hydrolysis of the calcic feldspar *anorthite*, on the other hand, leaves a residue of vermiculite-type clay, which is a weak acid, and the strong base $Ca(OH)_2$:

$$3\,Ca\,Al_2^{IV}\,Si_2O_8 + 6H_2O \rightleftharpoons 2H^+\,2[(Al\,Si_3)^{IV}Al_2^{VI}$$
$$O_{10}(OH)_2]^- + 3\,Ca(OH)_2 \qquad (9.2)$$

The rate of these reactions depends on temperature, and the availability of water, but especially on the area of crystal surface exposed and the rate of removal of the products. Since the products are weak acids and strong bases, the reactions are also favoured by high activities of H^+ ions.

Figure 9.2. Scheme for silicate mineral hydrolysis in the presence of CO_2-charged rainwater and humified organic residues.

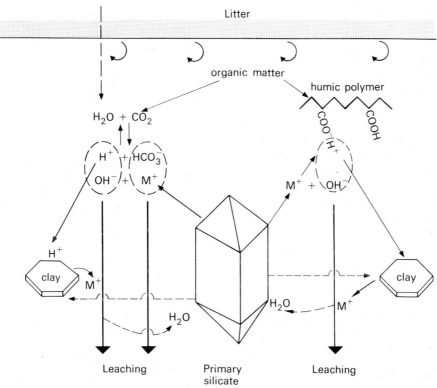

THE ROLE OF H$^+$ IONS

Equations (9.1) and (9.2) are specific examples of the general reaction:

$$[M^+ silicate^-] + H_2O \rightleftharpoons H^+[silicate\ secondary\ mineral]^-$$
$$+ M^+OH^- \qquad (9.3)$$

in which M$^+$ is an alkali or alkaline earth cation, or sometimes a transition element (Fe, Mn). Owing to the formation of the strong base MOH, nearly all the primary silicates have an alkaline reaction—the *abrasion* pH—when placed in CO$_2$-free distilled water (pH $\simeq 7$). However, reaction (9.3) proceeds more rapidly to the right when more H$^+$ ions are present to neutralize the OH$^-$ ions of the strong base: the main sources of this extra acidity in natural, well drained soils are

(a) CO$_2$ dissolved in the soil water, at partial pressures up to at least 0.01 bars (section 8.2) and

(b) Carboxyl and phenolic groups of humified organic matter (section 3.4).

The neutralization reactions and possible cation exchange reactions involving the clay mineral residues and the organic matter are illustrated in Figure 9.2.

As basic cations are removed from the soil by leaching, the clay minerals and organic matter gradually become saturated with H$^+$ ions. However, H-clays are unstable and slowly disintegrate from the crystal edges inwards to release Al and SiO$_2$, the silica going into solution as Si(OH)$_4$ and the Al being adsorbed as Al^{3+} by both clay minerals and humified organic matter (section 7.2). The subsequent slow hydrolysis of the Al^{3+}

produces more H$^+$ ions, the neutralization of which is achieved by further lattice decomposition so the resultant pH is higher than that of a pure H-clay or organic acid. Ultimately, clay minerals are completely destroyed to yield gibbsite, or goethite/hematite mixtures in soils high in ferromagnesian minerals. The change in mineralogy and pH with depth in a highly weathered soil on basalt (Table 9.2) illustrates this sequence of events.

Leaching

DIFFERENTIAL MOBILITIES

The mobility of an element during weathering reflects its solubility in water and the effect of pH on that solubility. Relative mobilities, represented by the Polynov series in Table 9.3, have been established from a

Table 9.3. The relative mobilities of rock constituents.

Constituent elements or compounds	Relative mobility*
Al$_2$O$_3$	0.02
Fe$_2$O$_3$	0.04
SiO$_2$	0.20
K	1.25
Mg	1.30
Na	2.40
Ca	3.00
SO$_4$	57

* Expressed relative to Cl taken as 100
(after Loughnan, 1969)

comparison of the composition of river waters with that of igneous rocks in the catchments from which they drain. The least mobile element is titanium, as anatase, which is insoluble at pH > 2.5. It is therefore used as a reference material to determine the relative gains or losses of other elements in the profile.

PERCOLATION DEPTH

Leaching of an element depends not only on its mobility but also on the rate of water percolation through the soil. In arid areas, even the most mobile constituents—mainly NaCl, and to a lesser extent the chlorides, sulphates and bicarbonates of Ca and Mg—tend to be retained and give rise to *saline soils* (section 9.6). As the

Table 9.2. Mineralogical changes in weathering basalt.

	Depth (cm)	pH (in water)	Kaolinite (%)	Other clay fraction minerals
Soil surface				
	10–20	6.5	15–25	H, Q, G
Decreasing	60–90	6.2	30–40	H, Q, G
intensity	120–150	6.1	40–50	H, G
of	180–210	6.0	45–55	H
weathering	270–300	5.7	55–65	H
	570–600	4.9	65–75	H
parent rock				

H = hematite; Q = quartz; G = gibbsite
(after Black and Waring, 1976)

climate becomes more humid, losses of salts and silica increase, and the soils become more highly leached except in the low-lying parts of the landscape where the drainage waters and salts accumulate and modify the course of soil formation (section 9.4). The effect of leaching is well illustrated by the steadily increasing depth to the carbonate layer in soils formed on calcareous loess in the U.S.A., along a line drawn from semiarid Colorado to humid Missouri (Figure 9.3).

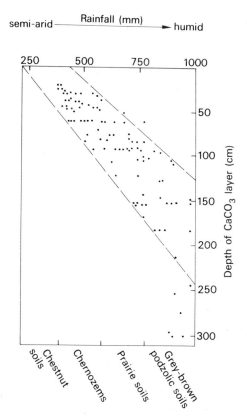

Figure 9.3. Leaching of $CaCO_3$ with increasing rainfall (after Jenny, 1941).

Cheluviation

In certain instances, the leaching of cations from the immobile oxides Al_2O_3 and Fe_2O_3 is enhanced by the formation of soluble organo-metal complexes. Aluminium forms an electrostatic-type complex with the carboxyl groups of humified organic matter, e.g.

$$\begin{matrix} COOH \\ R\text{---}COOH \\ COOH \end{matrix} + Al^{3+} \rightleftharpoons \begin{matrix} COO^- \\ R\text{---}COO^- Al^{3+} \\ COO \end{matrix} \quad (9.4)$$

Ferric iron, on the other hand, is reduced to ferrous iron at the expense of plant-derived reducing agents, with the concomitant formation of a soluble ferrous-organic complex:

$$\text{Organic reductant} + Fe^{3+} \rightleftharpoons Fe^{2+}\text{-organic ligand} \quad (9.5)$$

The most active compounds are the soluble polyphenols, such as epi- and D-catechin, that are leached in greatest concentration from the freshly fallen litter or canopies of conifers and heath plants. Although the complexes will form at pH 7, and in the presence of oxygen, they persist much longer at low pH and under anaerobic conditions. The formation and subsequent leaching of these complexes is called *cheluviation*. Polyphenols not involved in cheluviation are oxidatively polymerised to help form, together with the 'tanned' proteins and cellulose, the characteristic H-layer of mor humus found under coniferous forest and heath vegetation (section 3.3).

PODZOLIZATION

Podzolization involves the cheluviation of Al and Fe from insoluble oxides in the upper soil zone and their deposition, with organic matter, deeper in the profile. The removal of iron oxide and organic coatings renders the sand grains bleached in appearance, which is a feature of an Ea horizon. As the ferrous–organi ccomplex is leached more deeply it becomes less stable through a combination of the microbial breakdown of the organic ligand and the flocculating effect of higher concentrations of Al^{3+}, Ca^{2+} and Mg^{2+} in the soil solution near the weathering parent material. The iron precipitates as $Fe(OH)_3$, giving a uniform orange-red colour to the B horizon: the precipitated hydroxide adsorbs fresh ferrous–organic complexes, in which the iron subsequently oxidizes, and also acts as a coarse filter for humus and Al–organic complexes eluviated from the A horizon. The Bs horizon which forms is therefore usually overlain by a Bh horizon.

The end-product of podzolization is a *podzol*, the exact form of which is determined by the interplay of vegetation, drainage and parent material. Podzolization is favoured by permeable, siliceous parent material, a P/E ratio >1 for most of the year and slow rates of organic decomposition because of the type of litter or the acidity of the soil, or both. Two types of *podzol* profiles are shown in Figure 9.4.

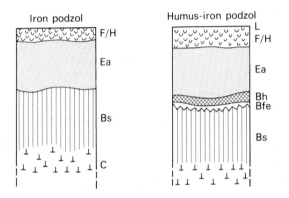

Figure 9.4. Diagrammatic profiles of an *iron podzol* and a *humus-iron podzol*.

LESSIVAGE

Where leaching is less severe, and the soluble organic complexes enjoy only a transient existence in the soil, the process of *lessivage* occurs. Duchaufour (1977) uses this term to describe the mechanical eluviation of clay and iron oxides from the A horizon without chemical alteration. The solution of iron during lessivage is probably qualitatively the same process as in podzolization, the difference being that the leaf leachates are less potent chelators, and the complexes formed are less stable. The removal of surface oxide films renders aggregates of clay domains and quartz grains unstable, whereupon the clay is washed progressively downwards in the percolating water to be deposited as clay coatings and gradually to form a Bt horizon.

Lessivage leads to a textural contrast between the Ea or Eb horizon and the Bt horizon. With increasing leaching ,acidity and longevity of the metal–organic complexes, soils pass through successive stages in the sequence:

sol brun lessivé → sol lessivé → sol lessivé podzolique → podzol

The first three soils in this series correspond to the *podzolic* or *podzolized* soils of American terminology (see Table 5.3). Essentially, the difference between a *podzol* and *sol lessivé* is that clay is destroyed rather than translocated in the *podzol*, and the intensity of cheluviation of Fe and to a lesser extent Al is greater in the *podzol*. The latter effect is illustrated in Table 9.4 by

Table 9.4. Free oxide ratios in the clay fractions of a typical *podzol* and *sol lessivé*.

	Podzol			Sol lessivé*	
Horizon	$\dfrac{SiO_2}{Al_2O_3}$	$\dfrac{Al_2O_3}{Fe_2O_3}$	Horizon	$\dfrac{SiO_2}{Al_2O_3}$	$\dfrac{Al_2O_3}{Fe_2O_3}$
Ea	3.22	5.72	Eb	3.15	3.98
Bh Bs	1.85	0.84	Bt1	3.05	3.66
Bs	1.83	2.40	Bt2	3.70	3.12
C	2.12	3.71	C	3.21	3.10

* Grey-brown podzolic soil (after Muir, 1961)

the higher Al_2O_3/Fe_2O_3 ratio in the clay fraction of the *podzol* eluvial horizon, and lower Al_2O_3/Fe_2O_3 and SiO_2/Al_2O_3 ratios in the B horizon of the *podzol* compared to the *sol lessivé*.

GLEYING

Under cool wet climates, the upper horizons of a *sol lessivé* may become waterlogged due to the impermeability of the Bt horizon. Even in a well drained soil, if the climate favours the formation of an O horizon or peat, the surface may remain saturated for most of the year. In other situations the whole soil may be waterlogged for all or part of the year due to high groundwater.

Collectively, soils that are influenced by waterlogging are called *hydromorphic* soils. The most important feature of such soils is *gleying* (section 8.4). True gley soils have uniformly blue-grey to blue-green colours and are depleted of iron due to its solution and removal as Fe^{2+} ions. The reduction of iron is primarily biological and requires both organic matter and microorganisms capable of respiring anaerobically. However, most of the iron exists as Fe^{2+}–organic complexes in solution or as a mixed precipitate of ferric and ferrous hydroxides, which is responsible for the characteristic gley colour. This precipitate forms as follows:

$$2Fe(OH)_3 + Fe^{2+} + 2OH^- \rightleftharpoons Fe_3(OH)_8 \text{ or } Fe_3O_4 . 4H_2O$$
$$\text{hydrated magnetite} \qquad (9.6)$$

Mottling is a secondary effect in gleys resulting from the reoxidation of iron in better aerated zones, especially around plant roots and in the larger pores. The orange-red mottles have a much higher Fe content than the surrounding blue-grey matrix. Thus, a C/G or B/G horizon frequently develops a coarse prismatic structure in which the ped interiors have a uniform gley colour and the exterior faces, pores and root channels have rusty mottling. This is typical of a *groundwater gley*. By contrast, in soils with impeded surface drainage, or those overlain by a thick O horizon, gleying is more subject to seasonal weather conditions and usually appears as blue-grey colours on the ped faces and normal oxidized colours in the ped interiors. This is typical of a *surfacewater gley*. Profile features of groundwater and surfacewater gleys are illustrated in Figure 9.5.

Figure 9.5. Profile features of typical groundwater gley and surfacewater gley soils (after Crompton, 1952).

9.3 FREELY DRAINED SOILS OF HUMID TEMPERATE REGIONS

The soils over substantial areas of Europe and North America have formed on comparatively uniform parent materials of glacial or periglacial origin for a similar period of time. On well drained sites of slight to moderate relief, two soil sequences are common—one on calcareous, the other on siliceous parent materials.

Soils on calcareous, clayey parent material
The general sequence of soil formation with the passage of time, and increased leaching, is shown in Figure 9.6. The plant succession advances from grassland to deciduous forest. In eastern Canada under deciduous forest the soil changes from a *brown forest soil* to *grey-brown podzolic soil*, but to a *grey wooded soil* under cooler conditions where conifers flourish. The major difference is that the *grey-brown podzolic* has a mull-like Ah horizon and the *grey wooded soil* a mor humus layer. Both soil types degenerate, with further

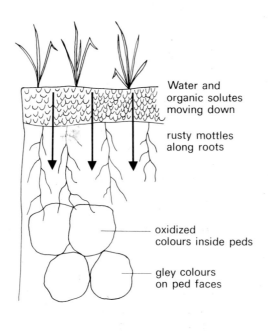

Water and organic solutes moving down

rusty mottles along roots

oxidized colours inside peds

gley colours on ped faces

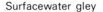
Surfacewater gley

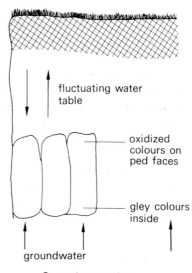

fluctuating water table

oxidized colours on ped faces

gley colours inside

groundwater

Groundwater gley

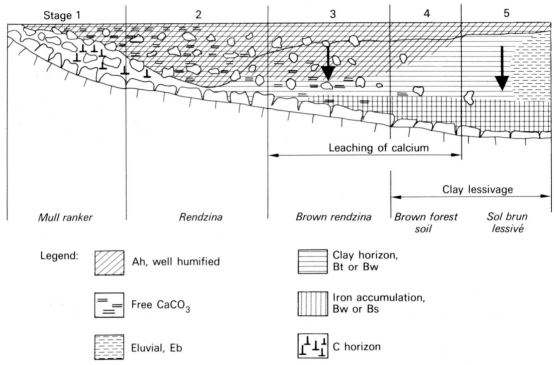

Figure 9.6. Soil maturity sequence on calcareous clayey parent material under a humid temperate climate (after Duchaufour, 1977).

leaching, into *brown podzolic soils* in which the whole profile becomes acidic and podzolization more obvious. The latter corresponds to the intermediate stages (4 and 5) of Duchaufour's maturity sequence on siliceous parent rock (Figure 9.7).

Soils on permeable, siliceous parent material
The early stages of this soil sequence, illustrated in Figure 9.7, feature clay lessivage, under a deciduous forest cover, leading to the formation of a *sol brun lessivé* and *sol lessivé*. With further time and leaching, the podzolizing influence strengthens and an *iron podzol* or *humus-iron podzol* forms in the surface of the deep permeable Eb horizon of the *sol lessivé*. Mackney (1961) has described the terminal stages (5 to 7) of this sequence under oakwoods on heterogeneous sands and gravels of the Bunter Beds in the English Midlands. His sandy brown earth, slightly podzolized, is a younger soil forming in the deep Eb horizon of a relic *sol lessivé*.

Soils on mixed siliceous and calcareous parent material
A sequence similar to Figure 9.7 occurs under beech-woods in the Chiltern Hills of England, where the original parent material consisted of some siliceous glacial drift mixed with soliflucted material from the Chalk that has, however, become acidic after prolonged weathering and leaching (Avery, 1958). As shown in Figure 9.8, the soils on the steepest slopes *A*, where erosion is active, are dominated by the Chalk rock and a typical *rendzina* is formed (*cf.* stage 2, Figure 9.6). On the intermediate slopes *B*, mantled by thin drift and solifluction deposits, immature *brown forest soils* are found. On the deeper drift and acid clay-with-flints of the plateaux and gentle slopes *C*, *sol brun lessivés* and *sol lessivés* occur. Some gleying occurs in the Bt horizons of *sol lessivés* in the valleys, while on the plateaux, mor humus accumulates and occasionally *micropodzols* are found in the Eb horizons of the *sol lessivés*.

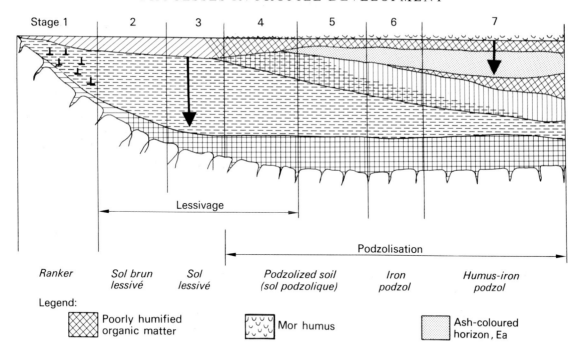

Figure 9.7. Soil maturity sequence on permeable, siliceous parent material under a humid temperate climate (after Duchaufour, 1977).

Figure 9.8. Parent material variations with relief in an old Chalk landscape of the Chiltern Hills (after Avery, 1958).

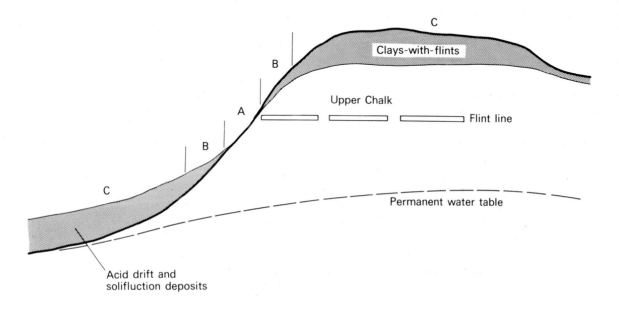

9.4 SOILS OF THE TROPICS

Pedogenic processes in the Tropics are not qualitatively different from those of temperate climes, but chemical and biochemical reactions proceed more rapidly and fluctuations in soil moisture are more extreme. Further, although much ravaged by erosion during the Pleistocene, many tropical land surfaces date from pre-Pleistocene times and some (notably in Africa and Australia) go back as far as the early Tertiary and late Mesozoic (60–100 m yr B.P.). Distinctive maturity sequences may be recognized on acidic and basic rocks, under rainfall regimes ranging from arid and semiarid to wet.

Acidic parent materials

Starting from the rudimentary soil forming on fragmented granite or quartzite—a *lithosol*, the course of soil formation under semiarid conditions proceeds as follows:

Lithosol → fersiallitic soil → ferrallitic soil

or under humid to wet conditions:

Lithosol → ferrisol → ferrallitic soil

The clay minerals formed are predominantly kaolinitic, with increasing degradation to sesquioxides and release of silica as the sequence advances. The *fersiallitic*, or ferruginous tropical soils, are not as highly leached as the ferrisols. They are rarely more than 250 cm deep and have medium levels of base saturation. The ferrisols, on the other hand, are leached and show lessivage. Structural B horizons are formed with coatings of sesquioxides and clay on the ped faces. *Ferrisols* are usually preserved, in an environment of intense weathering and leaching, on steeper slopes where the loss of surface soil is made good by progressively deeper weathering of the parent material.

The *ferrallitic soils* represent the mature stage in which only the most resistant minerals and immobile elements remain. Preserved by the flatness of the terrain, the *ferrallitic soils* may be very deep (up to 50 metres and more) and show little textural differentiation. Clay coatings in the Bw horizon have been destroyed, and the bulk of the clay minerals decomposed, leaving a residue high in quartz sand and sesquioxides. The *CEC* and base saturation are therefore low, and the soils very acid (pH ~ 4).

Basic parent materials

From a *lithosol* on basalt, for example, the soil maturity sequence proceeds under drier conditions through the stages:

Lithosol → eutrophic brown (chocolate) soils → vertisol

or under higher rainfall on well drained sites:

Lithosol → krasnozem → ferrallitic soil
↘ ferrisol ↗

The immature *brown* or *chocolate soils* have A horizons with deeply mixed organic matter and B horizons differentiated on the basis of colour, texture or structure. The clays are predominantly of the 2:1 type, yet the soils are permeable. The mature *vertisols*, however, have high clay contents (> 35 per cent) and poor permeability. The swelling capacity of the montmorillonite clay results in surface cracking, slickensides and gilgai microrelief (section 5.5).

Owing to greater leaching losses, *krasnozems* formed on basalt have predominantly kaolinitic clay with disseminated iron oxide giving the soils bright reddish-brown to red colours. The soils are, however, neutral to slightly acid in reaction, and not impoverished of plant nutrients. Only with continued weathering does the kaolinite break down to gibbsite, with silica release, and the soil becomes a *ferrallitic soil* high in sesquioxides.

LATERITE

Under hot and wet conditions, most parent materials are ultimately transformed to deep *ferrallitic soils*. Where the organic-rich A horizon of these old soils is stripped off by erosion, the exposed iron oxide zone becomes indurated by dehydration at high temperatures to form a geological structure called *laterite*. A typical laterite profile, illustrated in Figure 9.9a comprises:
(1) an uppermost *concretionary layer* of iron oxides, with secondary alumina, which may be massive if cemented, but loosely nodular or pisolitic* if uncemented.

---

* Resembling small dried peas in size and shape.

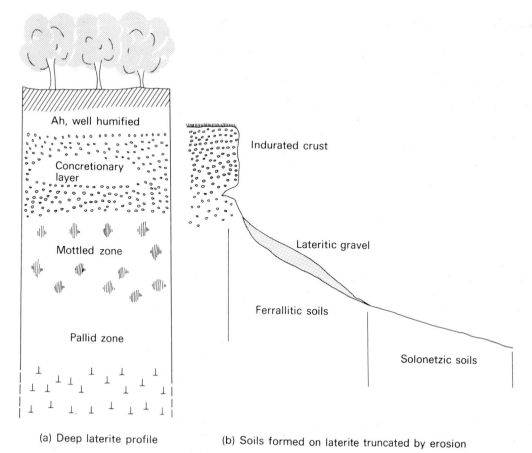

Figure 9.9. (a) A deep *laterite* profile. (b) Soil catena on a truncated *laterite*.

(2) An intermediate *mottled zone*—orange to red iron oxide deposits in a pale-grey matrix of kaolinite.

(3) Lowermost, a *pallid zone* of pale-grey to white kaolinite and bleached quartz grains, sometimes weakly cemented by secondary silica, and usually increasing in salt concentration towards the weathering zone.

Ancient laterites were much eroded and dissected during the Pleistocene and survive as prominent escarpments and crusts in the landscapes of tropical Africa, Australia and S.E. Asia. Truncation by erosion gives rise to a new weathering surface (Figure 9.9b) on which distinctive contemporary soils have formed—impoverished *ferrallitic* soils on the upper two laterite layers, and solonetz-type soils (section 9.6) in the pallid zone where salts had accumulated.

While layers 1 and 2 of the laterite profile represent the residuum of prolonged intense weathering, the pallid zone appears to form due to the reduction and solution of iron under anaerobic conditions in a zone

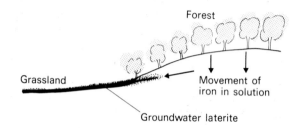

Figure 9.10. *Groundwater laterite* formation in a humid tropical environment.

of groundwater influence. The reduced iron moves in solution both laterally and vertically, the vertical movement following seasonal, or longer-term fluctuations in the height of the watertable. Reoxidation and precipitation of insoluble $Fe(OH)_3$ may in some cases account for the development of the overlying mottled zone. Nevertheless, iron moving laterally in percolating drainage water is eventually precipitated at some point in the landscape where better aeration supervenes, generally because of drier conditions, and the ironstone or *orstein* deposits typical of a *groundwater laterite* form (Figure 9.10).

9.5 HYDROMORPHIC SOILS

Soils showing some hydromorphic features are found in catenas in many parts of the world: the *vertisols* and *groundwater laterites* are two examples already cited. Nevertheless, hydromorphism is especially prevalent in humid temperate regions, such as the British Isles, where groundwater gleys and surfacewater gleys can be found on most parent materials. Two soil sequences found in the north and west of England and Wales serve to illustrate the point:

(1) At altitudes >300 m where the annual rainfall is >1250 mm, the high P/E ratio and low temperatures favour blanket peat formation (see below) even on well drained parent materials. A typical catena, proceeding downslope, consists of

Blanket peat → thin iron-pan soil → podzol or brown
*   podzolic soil*

as illustrated in Figure 9.11.

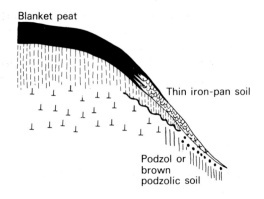

Figure 9.11. Catenary sequence on cool wet uplands in the west of England and Wales.

Gley colours predominate under the peat and to the base of the wet soil zone as the peat attenuates to a thin organic horizon downslope. It is roughly in this midslope area that the distinctive iron-pan—shiny black above and rusty-brown beneath—is found, as reduced iron in solution moves down into the better aerated lower horizons and is oxidized. This is an example of a *surfacewater gley*.

(2) At lower altitudes under an annual rainfall of 750 to 1000 mm, a catenary sequence similar to that of Figure 9.12 forms on slopes of gentle relief often overlying relatively impermeable glacial till. The sequence consists of:

Brown podzolic soil → peaty podzol → peaty gley soil
(Sol podzolique)

Figure 9.12. Catenary sequence on cool, imperfectly drained lowlands in the west of England and Wales.

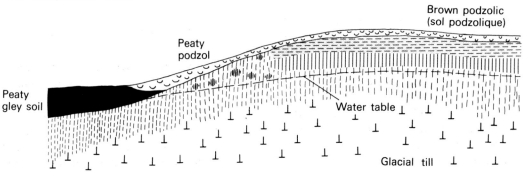

The better drained upslope soils show rusty mottling along root channels in an otherwise weakly gleyed Ea or Eb horizon underlying the Ah; but as the watertable rises downslope the whole profile becomes gleyed, apart from weak rusty mottling near the surface, and a *peaty gley soil* forms. This is an example of a *ground-water gley*.

Peat formation

A prerequisite of peat formation is a very slow rate of organic decomposition owing to a lack of oxygen under continuously wet conditions. This condition is achieved by various combinations of relief or climate (especially P/E ratio) with permeability and base status of the underlying rocks. The two main types of peat are described below.

BASIN OR TOPOGENOUS PEATS

Irrespective of climate, basin peats form where drainage water collects. If the water is neutral to alkaline, even though plant decomposition rates are substantial, the greater growth of sedges, grasses and trees produces an accumulation of well-humified remains in the form of *fen peat*. The ash content of this peat is high (10–50 per cent) due to the accession of colloidal mineral particles in the drainage waters. Large areas of fen peat are found, for example, in Florida, the Fens of England and in Botswana, the latter two being areas of relatively low rainfall.

BLANKET BOG (UPLAND OR CLIMATIC PEAT)

A P/E ratio >1 for all or part of the year and cool temperatures predispose to the formation of *blanket peats*—described as ombrogenous deposits—over much of the west of Ireland, Scotland and the Pennines of England, especially on acidic rocks. The vegetation of hardy grasses, Ericaceous plants and *Sphagnum* moss provides an acid litter, and since the only source of water is rain that is low in bases, the rate of plant decomposition is extremely slow and the humification imperfect. The ash content of the peat is usually < 5 per cent.

The formation of basin and fen peats began earlier in the post-Pleistocene period than did the blanket peats, many of which originated during the warm, wet Atlantic phase lasting from *c*. 7500 B.P. to 4000 B.P.

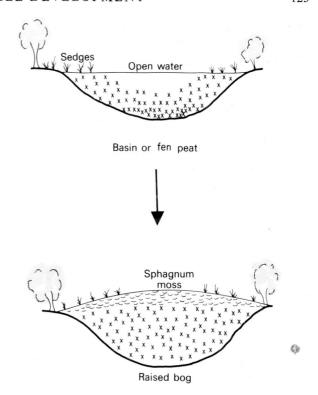

Basin or fen peat

Sphagnum moss

Raised bog

Figure 9.13. Transformation of a basin or fen peat to a raised bog.

Where present-day climate is favourable P/E ratio $>$) basin and fen peats are often transformed to *raised bog* (Figure 9.13). The build-up of sediment and peat in a wet hollow eventually raises the living vegetation above the level of the groundwater so that continued plant growth becomes dependent solely on rainwater. Slowly, acid-tolerant and hardy species such as heather, cotton-wood and *Sphagnum* invade and displace the species typical of the fen or basin peat.

9.6 SALT-AFFECTED SOILS

Soluble salts are derived initially from the weathering of primary rock minerals. Marine sediments, which contain *connate* salts, give rise to soils that are rich in salts and initially gleyed. Near the sea coast, even well-drained soils may accumulate salts that are brought down in the rainfall from consistent on-shore winds. This *cyclic salt*, as it is called, can also accrue from dust blown from arid areas where surface salts accumulate following the evaporation of drainage waters. The majority of salt-affected soils arise, however, because saline groundwater has come too near the soil surface.

A soil is *saline* if it has a salt concentration exceeding *c.* 2500 ppm, corresponding to an electrical conductivity of 4 mmhos cm^{-1} (determined in water just sufficient to saturate the soil sample). Naturally occurring saline soils, common in arid regions, are called *solonchaks* or *white alkali soils*: typically, they have a uniform dark-brown Ah horizon containing superficial salt efflores-

cences as well as interflorescences within the pores. The salts are brought up by capillary rise of the saline groundwater. Below the Ah horizon the soil is structure-less and frequently gleyed. Although there is some exchange of Na$^+$ for Ca^{2+} and Mg^{2+} on the clay surfaces, in accordance with the ratio law (section 7.2), the clay remains flocculated and the structure stable provided that a high salt concentration is maintained. The pH of such a soil lies between 7 and 8.3, depending on the occurrence of CaCO$_3$ and the partial pressure of CO$_2$ in the soil air.

Figure 9.14. *Solonchak-solonetz-solod* maturity sequence resulting from a falling watertable (after Kubiena, 1953).

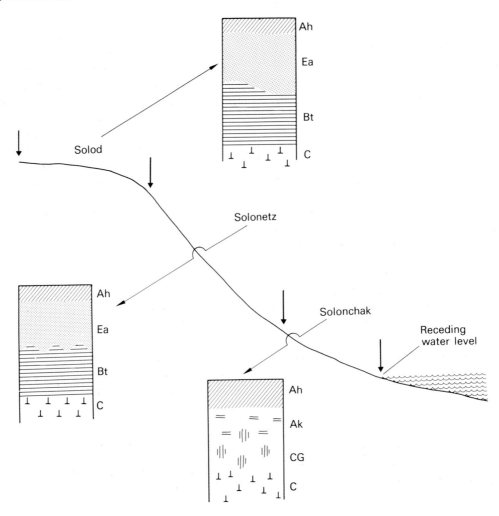

Solonchak–solonetz–solod soil sequence

Climatic change with rainfall increasing, or a lowering of the watertable, can have serious consequences for a saline soil, especially if the exchangeable Na^+ content exceeds 10 to 15 per cent of the *CEC*. The dilution of the soil solution as the salts are slowly eluviated causes the diffuse double layers at clay surfaces to expand, resulting in a pH rise and an increased swelling pressure within clay domains. Even dilution of a soil solution dominated by Ca and Mg from 1 M to 10^{-4} M would cause the pH to rise from 7 to 9 if the activity ratio $H^+/\sqrt{Ca^{2+}+Mg^{2+}}$ remained constant; with saline-sodium or *sodic* soils, pH values of 9–9.5 have been recorded. At such a high pH organic matter disperses, further weakening aggregate cohesion, and the soil passes towards the *black alkali* stage. The unstable Na-clay also disperses and is illuviated into the lower profile where it is flocculated by the higher salt concentration. The Bt horizon that develops is very poorly drained, often with signs of gleying, and has massive columnar peds that are capped with organic coatings (see Figure 4.4). The overall process of clay dispersion and illuviation is called *solonization*, and the soil formed is a *solonetz*.

Continued leaching of the solonetz leads to the complete removal of soluble salts, and the displacement of Na^+ ions from the clay surfaces by the main cation available—H^+, produced by the dissociation of carbonic acid. The pH of the surface soil falls below 7, and the H-clay in the Bt horizon slowly alters to an Al–Mg clay at the higher acidity, a change conducive to improved permeability of the Bt horizon compared with that of the *solonetz*. This, the final step in the degradation of a *solonchak*, is called *solodization*, and the resultant soil with its sandy, bleached E horizon and heavy Bt horizon is called a *solod* or *soloth*. The typical profiles of this maturity sequence are illustrated in Figure 9.14.

9.7 SUMMARY

A *soil profile* develops as the weathering front penetrates more deeply into the parent material and organic residues accumulate at the surface. The principal pedogenic processes operating are *mineral dissolution* and *hydrolysis, leaching, cheluviation, lessivage, gleying* and *salinization*, variations in the intensity of which with depth give rise to one or more *soil horizons*. Superimposed lithological discontinuities give rise to *soil layers*.

In temperate climates, with the passage of time, a soil passes through a *maturity sequence*, examples of which are given in Figures 9.6 and 9.7, as solutes are progressively leached out of the solum and organic matter accumulates. Furthermore, subtle changes in the influence of parent material, climate, organisms and relief can produce comparable profile variations in soils of the same age, as illustrated in Figures 9.11 and 9.12.

Pedogenesis in the Tropics proceeds faster than in temperate lands because of higher temperatures and a greater throughput of water. Compared with recently glaciated and periglacial areas of the Northern Hemisphere, much of the tropical land surface is very old so that soils have been forming on the same parent material for millions of years. A classic example is ancient *laterite* which survives, in a much eroded and dissected form, as striking relief features in tropical landscapes.

A new cycle of soil formation starts on the laterite profile when truncated by erosion. Similarly, when saline soils are leached due to climatic change or a receding watertable, *solonization* and ultimately *solodization* occur to produce the maturity sequence *solonchak* (saline soil) → *solonetz* → *solod*.

REFERENCES

AVERY B.W. (1958) A sequence of beechwood soils on the Chiltern Hills, England. *Journal of Soil Science* **9**, 210–224.

DUCHAUFOUR P. (1977) *Pédologie*. I. *Pedogenèse et Classification*, Masson, Paris.

HODGSON J.M. (1974) (Ed.) Soil Survey field handbook. *Soil Survey of England and Wales, Technical Monograph No. 5.*

MACKNEY D. (1961) A podzol development sequence in oakwoods and heath in central England. *Journal of Soil Science* **12**, 23–40.

FURTHER READING

BRIDGES E.M. (1970) *World Soils*. Cambridge University Press, Cambridge.

Davies R.I. (1971) Relation of polyphenols to decomposition of organic matter and to pedogenetic processes. *Soil Science* **111**, 80–85.

D'Hoore J.L. (1964) Explanatory monograph to soil map of Africa. *Commission for Technical Cooperation in Africa,* *Publication* **93**, pp. 1–126.

Kubiena W.L. (1953) *The Soils of Europe.* Murby, London.

Muir A. (1961) The podzol and podzolic soils. *Advances in Agronomy* **13**, 1–56.

Newbould P.J. (1958) Peat bogs. *New Biology* **26**, 89–104.

Part 3
Utilization of Soil

'Soils are the surface mineral and organic
formations, always more or less coloured by
humus, which constantly manifest themselves as a
result of the combined activity of the following
agencies; living and dead organisms (plants and
animals), parent material, climate and relief.'

V.V.Dokuchaev (1879), quoted by
J.S.Joffe in *Pedology*.

Chapter 10
Nutrient Cycling

10.1 NUTRIENTS FOR PLANT GROWTH

The essential elements

According to present knowledge, there are sixteen elements without which green plants can not grow normally and reproduce. On the basis of their concentration in plants, these *essential elements* are subdivided into the *macronutrients*

C, H, O, N, P, S, Ca, Mg, K and Cl

which occur at concentrations greater than 1000 ppm, and the *micronutrients*

Fe, Mn, Zn, Cu, B and Mo

which are generally less than 100 ppm in the plant. Carbon, H and O are only of passing interest, being supplied as CO_2 and H_2O which are abundant in the atmosphere and hydrosphere; likewise Cl which is abundant and very mobile as the Cl^- ion. Of the others, with the exception of N which comprises 79 per cent of the atmosphere, the major source is weathering minerals in the soil and parent material. The nutrient supplying power of a soil is a measure of its *fertility*, the full expression of which depends on diverse properties of the plant, its environment, and on management (section 11.2).

Plants absorb other elements, some of which are beneficial though not essential, such as Si (for sugar cane), Na (for sugar beet and mangolds) and Al (for many ferns). Cobalt is essential for symbiotic N_2 'fixation' in legumes and blue-green algae, while Se and I, although of no benefit to plants, are vital for the health and normal reproduction of animals.

Nutrient cycling

The flow of nutrients in the biosphere is continuous between three main compartments:

(a) an *inorganic store*, chiefly the soil;
(b) a *biomass store*, comprising living organisms above and below ground; and

(c) an inanimate *organic store*, on and in the soil, formed by the residues and excreta of living organisms.

The general cycle is illustrated in Figure 10.1. The importance of the various inputs and outputs, and the transformations undergone en route differs for each element, as does the partitioning between soil and non-soil. For example, contrast the behaviour of P where more than 90 per cent of the total element may be in the top 30 cm of soil, with that of K where the proportion is usually less than 50 per cent.

There are also differences in the way the elements are distributed within the soil. Elements that are released at depth by mineral weathering are absorbed by plant roots and transported to the shoots, to be deposited finally on the soil surface in litter and animal excreta. Accessions in rainfall add to the surface store. Nitrogen always accumulates in the organic-rich A horizon, the content declining gradually with depth; P is similar, the decline with depth being more abrupt because of the immobility of phosphate ions in the soil (Figure 10.2a). Sulphur, like N, accumulates in the surface of temperate soils, but not necessarily in tropical soils, owing to the mineralization of organic-S and the downward displacement of SO_4^-. Many of the latter soils typically show a bimodal S distribution, with organic S in the surface and an accumulation of SO_4-S in the subsoil (Figure 10.2b). The distribution of most cations—Fe^{3+}, Ca^{2+}, Mg^{2+}, Mn^{2+} and Zn^{2+} for example —correlates with clay accumulation, except in peaty soils.

THE INORGANIC STORE

With elements such as P, K and Ca plant roots tap only a small fraction of the soil's inorganic store during a single growth season. This fraction is withdrawn from the *available pool*, which is made up of

(a) Ions in the *soil solution*, and

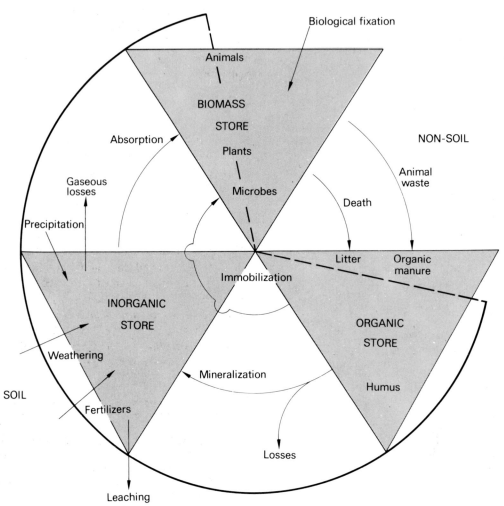

Figure 10.1. Fundamentals of a nutrient cycle.

(b) *Exchangeable ions* adsorbed by clay minerals and organic matter.

Less readily available ions are held in sparingly soluble compounds like gypsum, calcite and insoluble phosphates, or 'fixed' in non-exchangeable forms in clay lattices (K^+ and NH_4^+).

Inputs to the available pool occur by

(1) weathering of soil and rock minerals (section 5.2)
(2) precipitation and dust fallout
(3) mineralization of organic matter (section 3.1 *et seq.*)
(4) the application of organic manures, organic and inorganic fertilizers (section 11.1 and Chapter 12).

THE BIOMASS STORE

For an element such as K, which cycles rapidly through the soil–plant-animal system, the biomass store is a significant fraction of the total supply. For other elements the biomass store is important irrespective of how small it is, because the remnants of living organisms and their excreta are the precursors of manures and organic fertilizers.

THE ORGANIC STORE

The cycle of Figure 10.1 is completed by the decomposition of litter and organic manures by the soil

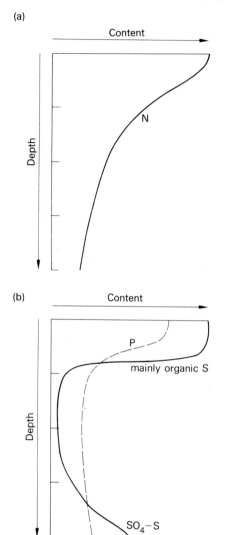

(a)

Content

Depth

N

(b)

Content

Depth

P

mainly organic S

SO_4–S

Figure 10.2 (a) and (b). Distribution of N, P and S in a soil profile.

the quantity of soil N varies from *c*. 1000 to 6000 kg ha^{-1}, almost all of which is in complex organic combination. Organic N is mineralized during microbial decomposition to release NH_4^+, according to the equation:

$$\frac{dN}{dt} = -kN \qquad (10.1)$$

where N = amount of organic N per unit soil volume (commonly 1 ha to 15 cm depth) and k is a rate constant varying between 0.01 and 0.06. For a time interval Δt of 1 year, the amount of N mineralized is given by kN, values of which range from 10 to 250 kg N ha^{-1}. Subject to limitations imposed by temperature, moisture, aeration and the effects of some plant species, NH_4^+ is oxidized to NO_3^- (section 8.3) and both ions comprise the pool of *mineral N* on which plants feed. Ammonium is held as an exchangeable cation which is readily displaced into the soil solution, and with NO_3^- moves to the roots in the water absorbed by the transpiring plant, a process of *mass flow*. Thus, all the mineral N (except for NH_4^+ 'fixed' in micaceous clay lattices) is available to the plant.

Table 10.1. N contents of selected agricultural crops.

	kg ha^{-1}
Wheat, 4 tonnes ha^{-1} (grain and straw)	80
Potatoes, 50 t ha^{-1}	180
Grass, 10 t ha^{-1} (dry matter)	250
Maize, 13 t ha^{-1} (grain and stover)	360
Sugar cane, 120 t ha^{-1}	500

mesofauna and micro-organisms, a topic discussed in Chapter 3. Specific aspects of mineralization, as they affect the availability to plants of individual nutrients, are discussed in the following sections.

10.2   THE PATHWAY OF NITROGEN

Availability of soil N to plants
Depending on the soil and environmental conditions,

In natural ecosystems, growth rates are low and the annual uptake of N is relatively small: Cole and others (1967) quote figures of 25–78 kg N ha^{-1} an^{-1} for a coniferous forest in North America. One of the highest figures is 248 kg N ha^{-1} an^{-1} estimated by Nye and Greenland (1960) for secondary rainforest in Ghana. Cultivated crops are much more demanding with N uptakes ranging from 100 to 500 kg N ha^{-1} (Table 10.1), so that the mineralizing capacity of the soil is often insufficient to maintain optimum growth. In this case, the magnitude of the gains and losses in the N cycle (Figure 10.3) are of great significance.

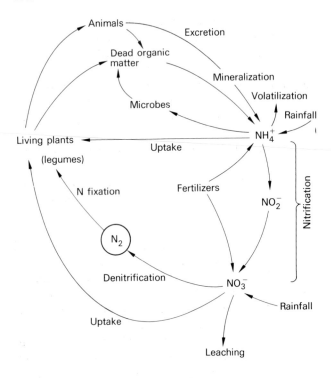

Figure 10.3. Transformations of N in the soil ecosystem.

Gains of soil N

PRECIPITATION

The concentration of N in rainfall depends on the industrial pollution of the air and its dust content, but usually ranges from < 1 to 5 ppm, mainly in the NH_4^+ form. The total N brought down also depends on the amount of rain falling, and figures round the world range between 1 and 30 kg N ha^{-1} an^{-1}. Most values are < 10 kg N so that the contribution of precipitation to the N cycle is small.

FERTILIZERS AND MANURES

Under natural vegetation, plant litter and organic manures provide the bulk of the nitrogen returning to the soil. In agriculture, however, where much of the crop is removed from the land, the return of N in chemical fertilizers or through the growth of legumes is necessary to bolster the soil N supplies for subsequent

crops. We shall now turn to the topic of biological N fixation.

BIOLOGICAL N FIXATION

A small minority of micro-organisms, either living free or in symbiotic association with plants, is exceptional in being able to reduce molecular N_2 to NH_3 and incorporate it into amino acids. This process, called *nitrogen fixation*, enables them to grow independently of mineral N. The annual turnover of biologically fixed N_2 in the biosphere is $\sim 10^8$ tonnes.

The splitting of the dinitrogen molecule ($N \equiv N$) and the reduction of each moiety to NH_3 is catalysed by the *nitrogenase* enzyme, which consists of two proteins—an Fe protein and a Mo–Fe protein, roughly in the ratio of 2:1. Reduction proceeds according to the reaction:

$$N_2 + 8H^+ + 6e + nMg.ATP \rightarrow 2NH_4^+ + nMg.ADP + nPO_4 \quad (10.2)$$

for which there are three prerequisites:
(a) a reductant of low redox potential (usually ferredoxin)
(b) an ATP generating system
(c) a low partial pressure of O_2 at the site of nitrogenase activity.

FREE-LIVING N_2 FIXERS

Regulation of N_2 fixation in aerobic organisms is complicated because the reduced nitrogenase must be protected from O_2, yet O_2 is required for the cell's respiration and oxidative phosphorylation. It is not surprising, therefore, that the ability to fix N_2 is not widespread among non-photosynthetic, aerobic organisms. Table 10.2 lists the more important of the free-living or non-symbiotic N_2 fixing organisms.

Table 10.2. The major free-living N_2 fixing organisms.

Bacteria	Aerobes	*Azotobacter vinelandii*
	Facultative anaerobes	*Azospirillum brazilense*
		Bacillus polymyxa; *Achromobacter* and *Pseudomonas* spp.
	Obligate anaerobes	*Clostridium pasteurianum*,
	Photosynthetic	*Desulphovibrio* spp.
Blue-green algae	Photosynthetic aerobes	*Rhodospirillum*, *Chlorobium* spp. *Nostoc*, *Anabaena* spp.

Apart from the oxygen status, factors such as pH, supply of organic substrates, mineral N concentration and the availability of trace elements, especially Cu, Mo and Co, influence the distribution and activity of N_2 fixing organisms in soil. *Azotobacter* and the blue-green algae are restricted to neutral and calcareous soils, but spore-forming anaerobes like *Clostridium* are tolerant of a wide range of pH. At high levels of mineral N, the N_2 fixers succumb to competition from more aggressive heterotrophic species. Thus, the rhizosphere of plant roots with its abundance of exudates of high C:N ratio is a favoured habitat for N_2 fixing organisms. In recent years, nitrogenase activity has been identified in the rhizosphere of many crops, such as wheat, maize, pasture grasses and sugar cane, a discovery made possible by the use of the *acetylene reduction technique*. When nitrogen is excluded from the nitrogenase site by an excess of acetylene, the latter is reduced to ethylene according to the reaction

$$C_2H_2 + 2H^+ + 2e \rightarrow C_2H_4 \qquad (10.3)$$

and the ethylene detected by gas chromatography.

Because of the high energy consumption of N_2 fixation (12–15 moles ATP per mole N_2 reduced), photosynthetic organisms tend to make the greatest contribution to non-symbiotic N_2 fixation. Estimates from field experiments (sometimes using ^{15}N as a tracer) suggest amounts from 0 to 100 kg N fixed ha^{-1} an^{-1}; estimates from farming systems are only 0–10 kg N ha^{-1} an^{-1}. The largest contribution probably occurs in waterlogged padi rice fields due to the activity of blue-green algae, and exceptionally in some tropical rain-forests where figures in excess of 100 kg N fixed ha^{-1} an^{-1} have been quoted.

SYMBIOTIC N_2 FIXATION

Symbiosis denotes the cohabitation of two unrelated organisms which mutually benefit from the close association. In the case of N_2 fixation, the invasion of roots of the host plant by a micro-organism (the endophyte) culminates in the formation of a *nodule* in which carbohydrate is supplied to the endophyte and amino acids formed from the reduced N are made available to the host.

There are three main symbiotic associations:
(1) Nodulating species of the family Leguminosae, in which the endophyte is a bacterium *Rhizobium*;
(2) Non-legumes, including the genera *Alnus* (alder), *Myrica* (bog-myrtle), *Elaeagnus* and *Casuarina*, in which the endophyte is usually an actinomycete;
(3) Lichens, being an association of a fungus and blue-green alga.

In terms of their numbers and widespread distribution, the legumes are by far the most important of the symbiotic N_2 fixers. A pictorial summary of the processes of root infection, nodule initiation and N_2 fixation in a legume root is presented in Figure 10.4.

The site of fixation is the bacteroid surface to which electrons are transported by a reduced co-enzyme and donated to $N \equiv N$ bound to nitrogenase. Synthesis of the nitrogenase is induced in the micro-organism under the favourable conditions existing within the host tissue. As in the aerobic free-living bacteria, regulation of the O_2 partial pressure on a microscale is essential for N_2 fixation to occur: this is achieved through the pigment *leghaemoglobin* (giving a characteristic pinkish-red colour to nodular tissue that is fixing N_2) which, because of its very high affinity for O_2, provides a means of transferring O_2 to the respiring bacteroids while keeping the free O_2 concentration in the membrane-bound sacs very low.

Other factors affecting N_2 fixation operate at any one of several points in the complex sequence of events: low pH and low Ca inhibit root hair infection and nodule initiation, at which stage also the competition between effective and non-effective bacterial strains is crucial. A non-effective strain produces many nodules which do not fix N_2, yet their presence inhibits nodulation by an effective strain. High mineral N, especially NO_3^-, suppresses nodulation while inadequate photosynthate and moisture stress reduce a nodulated root's fixation capacity. Nutritionally, nodulated legumes need more P, Mo and Cu than unnodulated legumes, and have a unique requirement for Co. Further details are given in recent reviews, e.g. Quispel (1974).

Because of environmental constraints, and inherent differences in host-bacterial strain performance, the quantity of N fixed is variable. Estimates for several

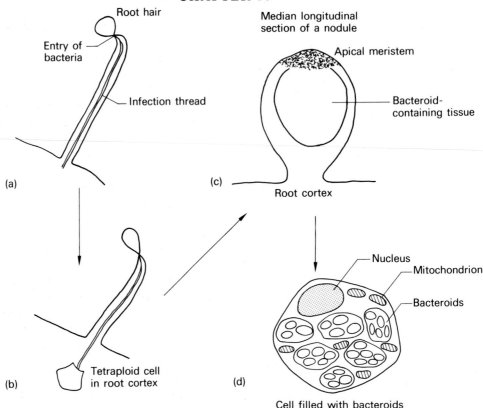

Figure 10.4. Sequence in nodule formation on a legume root. (a) Bacteria invade a root hair through an invagination of the cell wall—the infection thread. (b) Tetraploid cell penetrated—this cell and adjacent cells induced to divide and differentiate to form a nodule. (c) Bacteria released, divide once or twice and enlarge to form bacteroids which respire but cannot reproduce. (d) Groups of bacteroids enclosed with a membrane and N_2 fixation begins. (after Nutman, 1965).

temperate and tropical legumes are given in Table 10.3, which shows that under favourable conditions, the contribution of legumes to soil N can be appreciable. The legume N is made available by the sloughing-off of nodules, through the excreta of grazing animals, or by decomposition of the plant when it is ploughed into the soil.

Losses of soil N

CROP REMOVAL

Much of the crop N is removed in the harvested material, except under grazing conditions where some 85 per cent of the N is returned in animal excreta. The fate of cereal crop residues is important, for the straw from a 4 tonne ha^{-1} wheat crop contains some 20–25 kg N. The straw is baled and used as animal bedding, eventually forming farmyard manure, or burnt, when most of the N is volatilized, or occasionally ploughed back into the soil.

LEACHING

Nitrogen in the NO_3^- form is very vulnerable to leaching. Soil nitrate levels fluctuate markedly due to the interaction of temperature, pH, moisture, aeration and

Table 10.3. Estimation of N_2 fixation by legumes as measured in kg ha^{-1} an^{-1}.

	Temperate species		Tropical and subtropical species	
Clovers			Grazed grass-legume pastures:	
(*Trifolium* spp)	55–600		*Stylosanthes*	
Lucerne			*humilis*	10–30
(*Medicago sativa*)	55–400		*Macroptilium*	
Soyabeans			*atropurpureum*	44–129
(*Glycine max*)	90–200		Grain and forage legumes:	
			Beans	
			(*Phaseolus vulgaris*)	64
Peas (*Pisum* spp)	33–160		Pigeon pea	
			(*Cajanus cajan*)	97–152
Median between	100–200		Median between	50–100

species effects on nitrification. The following trends are noteworthy:

(a) *Temperature*. In temperate regions, low winter temperatures retard nitrification, which then attains a peak in spring and early summer when NO_3^- concentrations of 40–60 ppm N may be found in surface soils under bare fallow. Higher values are recorded in warmer, drier soils because of capillary rise and minimal leaching.

(b) *Moisture*. In warm climates of little seasonal temperature variation, soil mineral N levels follow the sequence of wet and dry seasons, as has been observed

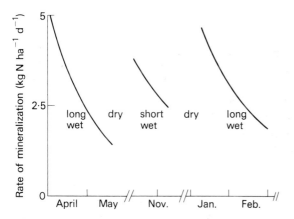

Figure 10.5. Seasonal changes in mineral N in an East African soil (after Birch, 1958).

in East Africa. Following partial sterilization of the soil during the long dry season, there is a flush of mineralization and nitrification at the onset of the rains leading to high NO_3^- levels which decline during the subsequent wet period (Figure 10.5).

(c) *Natural grassland and forest communities*. The mineral N found in grassland and forest soils is predominantly NH_4^+, which suggests that nitrification is suppressed, possibly due to the secretion of inhibitory compounds by the roots of certain species. Normally the input of N in rain roughly balances the loss by leaching in natural plant communities; clear-felling a forest, however, accelerates the rate of nitrification and in the absence of plants to absorb the NO_3^- leads to heavy N losses in the runoff (section 10.4).

(d) *Soil physical properties*. Soil texture and structure markedly influence NO_3^- leaching. In sandy soils, with a low volume of water held in large pores of high hydraulic conductivity, NO_3^- is rapidly leached: for a soil of $\theta = 0.2$ at field capacity, 100 mm of rain may displace NO_3^- downwards to a depth of $100/0.2 = 500$ mm. Conversely, in a well structured clay soil, where the pore water volume is large ($\theta \sim 0.4$), 100 mm of rain may displace NO_3^- only 250 mm. However, if most of the water and NO_3^- is held in fine intraped pores which do not allow ready mixing of the static and mobile water volumes, the *average* displacement of NO_3^- is less than the expected figure.

GASEOUS LOSSES

Animal excreta containing urea or uric acid (in the case of poultry) is susceptible to volatile losses of N. Catalysed by the enzyme urease (of microbial origin), urea is hydrolysed to ammonium carbonate which rapidly decomposes according to the reaction:

$$(NH_2)_2CO + 2H_2O \rightleftharpoons (NH_4)_2CO_3 \rightleftharpoons 2NH_3 + CO_2 + H_2O \quad (10.4)$$

At high temperatures and under alkaline conditions, ammonia gas is lost from solution. Volatilization of NH_3 also occurs from ammonium and urea fertilizers under unfavourable conditions (section 12.2).

Nitrogen losses by denitrification occur by two possible mechanisms—chemo-denitrification and biological denitrification, of which the latter is by far the more important.

(a) *Chemo-denitrification.* Losses of N have been observed, even from well aerated soils, when NO_2^- has accumulated following heavy applications of ammonium or urea fertilizers. The losses are greatest in acid and neutral soils and appear to involve the reaction of nitrous acid (HNO_2) with the soil organic matter to release gaseous N compounds. The most likely active organic components are phenolic compounds which are known to react with HNO_2 to form nitrosophenols that are decomposed under mildly acid conditions in the presence of excess HNO_2 to N_2O and N_2.

(b) *Biological denitrification.* As discussed in section 8.4, biological denitrification proceeds in anaerobic soil culminating in the release of N_2O and N_2 to the atmosphere. Quantification of the N loss is very difficult, because although N_2O concentrations in the soil air can be measured accurately by gas chromatography, N appearing as N_2 gas can only be distinguished from background N_2 gas if the original N substrate is labelled with ^{15}N. The rate of denitrification is also very variable, being dependent on organic substrate availability, temperature, pH and the distribution of anoxic zones in the soil. It is generally insignificant at pH < 6 and temperatures $< 10°C$. Deep leaching of NO_3^- to a poorly aerated subsoil during winter does not necessarily predispose to denitrification because

(i) the temperature is low, and

(ii) organic substrate is insufficient for rapid microbial growth.

Denitrification losses range from < 1 to 20 per cent of the annual turnover of mineral N.

10.3 PHOSPHORUS AND SULPHUR

Plant requirements

Crop analyses indicate that comparable amounts of P and S are removed in harvested products each year (Table 10.4); nevertheless, the two elements differ markedly in the balance of inputs and outputs in the soil–plant cycle, and to a lesser extent in the transformations which they undergo in the soil.

The soil reserve

RAINFALL INPUT

P concentrations in rain are very low, the amount

Table 10.4. Amounts of P and S in crop products.

	P	S
	(kg ha^{-1})	
Barley, 4 t ha^{-1} (grain and straw)	12	15
Potatoes, 50 t ha^{-1}	25	20
Grass, 10 t ha^{-1} (dry matter)	30	15
Maize, 13 t ha^{-1} (grain and stover)	50	50
Kale, 50 t ha^{-1} (fresh weight)	25	100

(after Cooke, 1975, and others)

deposited ranging from 0.2 to 0.5 kg ha^{-1} an^{-1}. Amounts of S are much higher, varying from 1–100 kg ha^{-1} an^{-1}, and rarely less than 15–30 kg in industrialized countries. The major source of this S is the burning of fossil fuels, with small contributions from sea spray near coasts, and the release of H_2S from marine sediments. The direct absorption of SO_2 by soils and plants (via stomata), which is called *dry deposition*, can provide 1/4 to 1/3 of a plant's S needs.

FORMS OF SULPHUR

Soil S is derived originally from sulphidic minerals in rocks which are oxidized to sulphate on weathering. Clays and shales of marine origin are frequently high in sulphides, and recently formed soils on marine muds may have SO_4-S contents up to 50,000 kg ha^{-1}. Normally, soil S is predominantly in an organic form and ranges between 200 and 2000 kg ha^{-1}.

Organic S is released by microbial decomposition, depending on the C:S ratio of the residues, which lies between 50 and 150 for well humified organic matter. Sulphur appears to be stabilized in humus in much the same way as N, and indeed the N:S ratio (between 6 and 10) is much less variable than the C:S ratio. The sulphate released is only weakly adsorbed at positively charged sites, in competition with phosphate and organic anions, and so is prone to leaching.

FORMS OF PHOSPHORUS

Soil P contents range from 500 to 2500 kg ha^{-1}, of which 15 to 70 per cent may exist in the strongly adsorbed or insoluble inorganic forms described in section 7.3; the remainder occurs as organic P which is the major source of P for the soil micro-organisms and

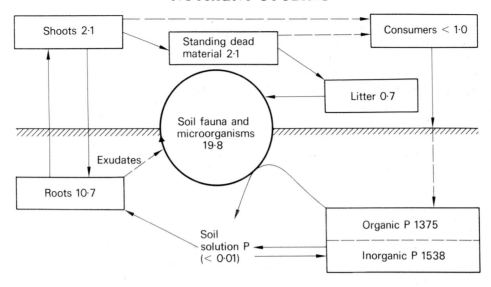

(Figures in kg ha^{-1} 30 cm depth)

Figure 10.6. P cycles in a native prairie grassland in Western Canada (after Halm and others, 1972).

mesofauna. The phosphate of nucleic acids and nucleotides is rapidly mineralized, but when the C:P ratio rises above *c.* 200, the P is immobilized by the microorganisms, especially bacteria, which have a relatively high P requirement (1.5–2.5 per cent P by dry weight compared to 0.05–0.5 for plants). Bacterial phosphate residues comprise mainly the insoluble Ca, Fe and Al salts of inositol hexaphosphate which are called *phytates.* Thus, in closed ecosystems with insignificant P inputs, the soil organisms are highly competitive with higher plants for P, as illustrated by the P cycle within a natural grassland (Figure 10.6).

Not only do micro-organisms mineralize organic P, but some groups secrete organic acids, such as α-ketogluconic acid, which attack insoluble Ca phosphates and release the phosphate. Species of *Aspergillus*, *Arthrobacter*, *Pseudomonas* and *Achromobacter*—which are abundant in the rhizosphere of plants—have this ability. However, most of the P is absorbed by microorganisms themselves in the intensely competitive environment near the root surface, where their growth is stimulated by the greater availability of carbon substrates than in the bulk soil.

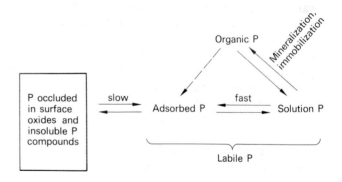

Figure 10.7. Phosphate transformations in soil.

PHOSPHATE AVAILABILITY

Orthophosphate ($H_2PO_4^-$ and $HPO_4^=$) that is released by mineralization is rapidly adsorbed by the soil particles where its availability steadily declines with time, a process called phosphate '*fixation*'. Phosphate on surfaces that can be readily desorbed, plus phosphate in solution, is called *labile P* to distinguish it from the P held in insoluble compounds, or in organic matter, which is *non-labile.* Transformations between the different forms of P in soil are represented in Figure 10.7.

Soil pH, clay and sesquioxide content and exchangeable Al^{3+} all influence P availability. Where the effects of the free Fe and Al oxides are predominant, or in fertilized acid soils where compounds of composition $FePO_4 . nH_2O$ and $AlPO_4 . nH_2O$ exist and dissolve incongruently to release more P than Fe or Al, P in the soil solution increases with pH rise. By contrast, in phosphate-rich alkaline and calcareous soils, insoluble Ca phosphates such as hydroxyapatite and octacalcium phosphate comprise the bulk of the non-labile P, the solubility of which improves as the pH falls. Consequently, the availability of P in such soils is greatest between pH 6 and 7.

However, acid soils of low P status which retain a surplus of clay minerals relative to sesquioxides show a contrary trend in P availability with pH change. Between pH 4.5 and 6, exchangeable Al^{3+} hydrolyses to form hydroxyaluminium ions which act as sites for P adsorption because of their residual positive charge. These ions also polymerize readily so that a hydroxyaluminium complex with $H_2PO_4^-$ ions *occluded* in its structure forms on the surface of the clay mineral, rendering a minimum in P concentration over the pH range 5–6.5. Above pH 7, Al begins to dissolve as the aluminate anion and P is released.

MYCORRHIZAS

Because of low P concentrations in soil solutions and fierce competition from micro-organisms, many plants have evolved mechanisms to enhance the absorption of P when it is in short supply; one of the most important of these is the mycorrhiza, a symbiotic association of fungus and root. There are two types of mycorrhiza:

(a) *The ectotrophic form.* This is primarily associated with trees (conifers and beech), the fungal mycelium growing externally and forming a sheath some 0.05 mm thick around the root cylinder. The external hyphae afford a small increase in the soil volume exploited by the root, but the major benefit derives from the fungus's ability to store P in its tissues, to be released to the host in times of P deficiency, and to the greater longevity of infected roots which continue to absorb P long after non-mycorrhizal roots have ceased active uptake. In return, the fungus derives carbon substrates and other growth requirements from the host.

(b) *The endotrophic form.* This form, in which the

Fig. 10.8. Diagram of a vesicular–arbuscular mycorrhiza (after Nicholson, 1967).

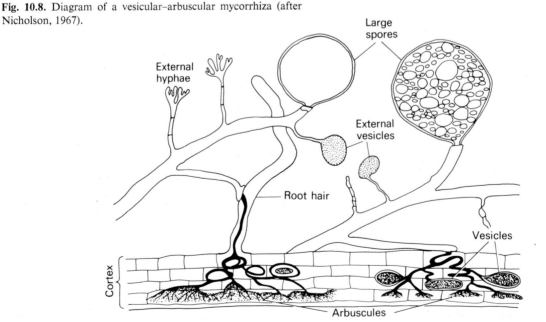

endophyte is a Phycomycete of the genus *Endogone*, is commonly called a *vesicular–arbuscular* (V–A) mycorrhiza. The fungal spores are present in most soils, but only germinate when a suitable host root is near: if the young mycelium does not penetrate a host root, it dies. However, once established in the root (and the host-fungal relationship is not very strain-specific), the hyphae ramify widely outside the host tissue.

Two structures are characteristic of the internal mycelium—firstly, the *vesicles* which are temporary storage bodies in the intercellular spaces; secondly, the *arbuscules* which are branched structures growing intracellularly and which probably release solutes into the cell lumen (Figure 10.8). V–A mycorrhizas are found in nearly all the angiosperms and in many conifers, ferns and lower plants. The incidence of infection is especially high in impoverished soils, where the mycorrhizal state confers a distinct advantage in P uptake, because the external hyphae add up to 50 per cent to the length of absorbing root and provide a pathway for the rapid transport of phosphate across the depletion zone around the root. The benefit is most notable in plants like onion and citrus which have no or very few root hairs.

Although V–A mycorrhizas are of great importance in the cycling of P in natural ecosystems, and for the establishment of pioneer species in harsh habitats such as on sand dunes and mine spoil, it is found that with cultivated crops the incidence of infection declines markedly as P fertilizers are applied, so that a growth stimulus due to mycorrhizas is not normally observed in agricultural soils.

LEACHING LOSSES

With the exception of very sandy soils, and other situations described in section 12.3, phosphate losses by leaching are very small, usually < 1 kg ha^{-1} an^{-1}.

Sulphate forms sparingly soluble gypsum in arid region soils, but is leached from soils of humid regions, except those high in sesquioxides and of low pH, where it is retained at depth (see Figure 10.2b). Data obtained for three sites on neutral and calcareous soils in southern England illustrate the contrast between leaching losses of sulphate and phosphate (Table 10.5).

Table 10.5. P and S losses in drainage from neutral and calcareous soils in southern England.

	PO$_4$-P	SO$_4$-S
	(kg ha^{-1} an^{-1})	
Woburn—neutral, sandy soil over Oxford Clay	0.05	100
Saxmundham—sandy, calcareous boulder clay	0.12	86
Wytham—brown calcareous soil over Oxford Clay	0.13	78

(after Williams, 1970, and others)

10.4 POTASSIUM, CALCIUM AND MAGNESIUM

Plant requirements

Agricultural crops remove annually between 5 and 25 kg Mg ha^{-1}, and exceptionally up to 50–60 kg ha^{-1} for high yields of maize or oil palm; Ca requirements range from 10 to 100 kg ha^{-1} and K even higher at 100 to 300 kg ha^{-1}.

Estimates of the rate of cation uptake by perennial vegetation, especially forests, are sparse because of the

Table 10.6(a). Annual return (=uptake) of K, Ca and Mg by a 40-year-old forest in West Africa.

	K	Ca	Mg
		(kg ha^{-1})	
Litter fall	68	206	45
Timber fall	5	82	8
Leaf wash	220	29	18
Total	293	317	71

(after Nye and Greenland, 1960)

Table 10.6(b). Annual return (=uptake) of K and Ca by 36-year-old Douglas fir.

	K	Ca
	(kg ha^{-1})	
Litter fall	3	11
Stemflow	1	1
Leaf wash	11	4
Total	15	16

(after Cole and others, 1967)

difficulty of measuring the amount and composition of the annual growth increment. By assuming that in a mature forest, the rate of nutrient return balances the rate of uptake from the soil, Nye and Greenland (1960) obtained the estimates of cation uptake shown in Table 10.6a. Much lower figures for K and Ca uptake were obtained by Cole and others (1967) for a 36 year old Douglas fir plantation in Washington, U.S.A. (Table 10.6b).

Soil reserves

The pool of exchangeable cations Ca^{2+}, Mg^{2+} and K^+ provides the immediate source of these elements for plants. Insoluble reserves may occur as calcium and magnesium carbonates, potash feldspars and interlayer K^+ in micaceous clays. Total reserves are therefore very variable, reflecting the conditions of soil formation, but exchangeable Ca^{2+} values usually lie between 1000 and 5000 kg ha^{-1}, Mg^{2+} between 500 and 2000 kg ha^{+1} and $K^+ \sim 1000$ kg ha^{-1}.

RAINFALL INPUT

Amounts of K and Ca in all forms of precipitation are comparable and lie between 1 and 20 kg ha^{-1} an^{-1}; Mg can be much higher, especially on sea coasts (for ocean water is rich in Mg), amounting to as much as 150 kg ha^{-1} an^{-1}. Thus, Mg inputs roughly balance combined losses through plant removal and leaching, but not so for K and Ca. A striking feature of K input to the soil is the magnitude of its leaching from leaves, amply demonstrated by the figures for the tropical species and Douglas fir in Table 10.6. This process is one of the main reasons for the rapid cycling of K in natural ecosystems in which plants and their residues can accumulate from 40 to 60 per cent of the total labile K which is within the root zone.

MINERAL WEATHERING

Apart from some recent glacial materials, the reserves of weatherable minerals are usually small in the top 30 cm of soil where 80–90 per cent of the plant roots are located. Nevertheless, the slow release of nutrients from the 'geologic reserve' at greater depth is often crucial for the stability of natural plant communities wherein elemental losses by leaching and erosion occasionally exceed atmospheric inputs. For example,

West African figures suggested that 58, 64 and 14 kg ha^{-1} of K, Ca and Mg were pumped up annually from the subsoil by deep tree roots.

LEACHING LOSSES

Cation leaching is accelerated in cultivated soils, where nitrification occurs, for two reasons (a) nitrification is an *acidifying* reaction (see equation 8.10), so that H^+ ions are made available to displace exchangeable Ca^{2+}, Mg^{2+} and K^+; (b) the increase in NO_3^- concentration must be balanced by an increased cation concentration in the water percolating through the soil. For agricultural soils in Britain, Ca leaching normally ranges between 160 and 320 kg ha^{-1} an^{-1}.

Conversely, suppression of nitrification in soils under natural grassland or forest tends to reduce cation losses by leaching. For example, in a deciduous forest catchment in New Hampshire, the net leaching loss of K, Ca and Mg from an acid podzolic soil was 2, 9 and 3 kg ha^{-1} an^{-1}, respectively, and there was a small gain of 2 kg N ha^{-1} an^{-1} (Likens *et al.*, 1970). However, in two years after the forest was felled, but neither burnt nor the timber removed, some 120 kg N ha^{-1} was lost annually (mainly as NO_3^-) and K, Ca, Mg losses rose to 36, 90 and 18 kg ha^{-1} an^{-1} respectively. Most of this increase was due to the mineralization of litter and soil organic matter exposed after the forest clearance, and accelerated nitrification which depressed the pH of the drainage water from 5.1 to 4.3.

10.5 TRACE ELEMENTS

Soil reserves

Those elements whose total concentration in the soil is invariably < 1000 ppm are called *trace elements*: they fall into two categories:

(a) The *micronutrients* Cu, Zn, Mn, B and Mo which are beneficial to plants at normal concentrations in the plant (ranging from 0.1 ppm for Mo to 100 ppm for Mn), but are toxic at higher concentrations.

(b) Other elements, notably Be, Bi, Cd, Cr, Pb and Ni which are of no benefit and are toxic at levels greater than a few ppm. Iron is the one micronutrient which is not strictly a trace element.

INFLUENCE OF PARENT MATERIAL

Trace elements are largely bound in mineral lattices, to be released only by weathering, so that the type of parent material determines to a large extent the abundance of the individual elements in the soil. Iron, Cu, Mn, Zn, Co, Mo, Ni and Pb occur in ferrogmagnesian minerals, common in ultrabasic and basic igneous rocks; Fe, Mn, and Mo also occur as insoluble oxides (sometimes with Co coprecipitated in the MnO_2), and Zn, Fe and Cu as equally insoluble sulphides in sedimentary rocks. Boron occurs as the resistant mineral *tourmaline* in acid igneous rocks. Trace element contents of soils formed on parent rocks of different basicity are presented in Table 10.7.

Table 10.7. Trace element contents of soils on parent materials of different rock type.

	Serpentine	Andesite	Granite	Sandstone
	←————————Increasing basicity————————→			
	(kg ha⁻¹)			
Co	160	16	< 4	< 4
Ni	1600	20	20	30
Cr	6000	120	10	60
Mo	2	< 2	< 2	< 2
Cu	40	20	< 20	< 20
Mn	6000	1600	1400	400

(after Mitchell, 1970)

Availability to plants

WATER SOLUBLE IONS AND ADSORBED FORMS

The processes involved in the turnover of a trace element in the soil–plant system are summarized in Figure 10.9.

Molybdenum, As and Se occur as anions in the soil solution and B as the undissociated acid $B(OH)_3$. In the absence of complexing agents, Fe, Cu, Zn, Mn, Co, Ni and Pb occur as cations; Fe, Cu, Zn and Mn undergoing hydrolysis at pH values between 2 and 10. Plant availability of these cations is better correlated with their solubility in acetic acid at pH 2.5 than with the small amounts extracted in neutral ammonium acetate (the normal extractant for exchangeable cations). Because of hydrolysis and the eventual precipitation of the insoluble hydroxides, the cation availability decreases as the pH rises. On the other hand, anions are least available at low pH due to adsorption by sesquioxides and become more available with pH rise. Boron is little adsorbed below pH 8, at which point the weak boric acid begins to dissociate and borate is adsorbed on oxide surfaces.

Trace elements in the form of
(a) insoluble organo-metallic complexes,
(b) precipitated oxides or insoluble salts, and
(c) resistant primary minerals

Figure 10.9. Trace element (E) equilibria in the soil (after Hodgson, 1963).

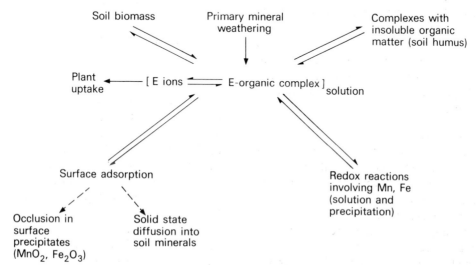

are of lesser importance for plant uptake, except in the case of poorly drained soils discussed below.

ORGANIC COMPLEXING OF TRACE ELEMENTS

Much of the trace element content in the soil solution, especially in Ah horizons, is present as organo-metal complexes of the kind described in section 3.4. For the divalent cations, the most stable complexes are formed with Cu followed by Mn, Zn and Ni; for the trivalent cations complexes of Fe are more stable than Co. The complexed ions behave as anions of a low charge/mass ratio in the soil solution. Despite this solubilizing effect, trace elements are also immobilized by adsorption onto, or complex formation with, insoluble humic compounds, which accounts for the incidence of Cu deficiency, for example, on freely drained peaty soils.

Trace elements in organic combination are estimated by extraction in ethylene diamine tetraacetic acid (EDTA), a synthetic chelating agent which forms very stable complexes with most trace elements. Use is made of such chelators in increasing the availability of Fe, Mn, Cu and Zn in alkaline soils (deficiency of Fe in such cases giving rise to plant symptoms of 'lime-induced chlorosis*'). The stability of the metal chelate depends on the solubility product of the metal hydroxide, the soil pH and the affinity of the chelator for the metal in question relative to other cations in abundance (e.g. Ca). Figure 10.10 shows that when the insoluble hydroxide maintains a lower metal ion activity in solution than the metal–chelate complex, formation of the hydroxide is favoured—the change over occurs at pH 9 for Fe EDTA in non-calcareous soils, and at pH ∼8 in calcareous soils. Currently, chelating agents which have a greater affinity for Fe relative to Ca than EDTA are used to correct Fe deficiency.

THE EFFECT OF POOR DRAINAGE

The manganese and iron of insoluble compounds is mobilized in waterlogged soils as Mn^{2+} and Fe^{2+} ions, as discussed in section 8.4. In addition, the solubility of Co, Ni, Mo and Cu (elements not reduced at the redox potentials attained in soils) may be increased

* chlorosis—the yellowing of leaves due to the lack of chlorophyll.

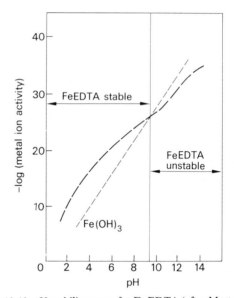

Figure 10.10. pH stability range for Fe EDTA (after Martell, 1957).

under waterlogged conditions due to the accelerated weathering of ferromagnesian minerals (Table 10.8). The dissolution of manganese oxides can also release coprecipitated Co into the soil solution.

Table 10.8. The effect of drainage on trace element solubility in soil.

| | Extractable* element content | | | |
	Co	Ni	Mo	Cu
	(ppm soil)			
Freely drained	1.3	1.3	0.06	2.6
Poorly drained	1.9	3.4	0.19	6.6

* extracted in CH_3COOH or EDTA

LEACHING LOSSES

Because of the precipitation of insoluble compounds, strong adsorption on mineral surfaces and complexing by soil humus, trace element losses by leaching are very small, except for B, and Fe and Mn in some gleyed soils. Nevertheless, without significant atmospheric inputs, cumulative losses from very old soils eventually lead to widespread micronutrient deficiencies, as occur in the coastal areas of eastern and southern Australia.

10.6 SUMMARY

Gains and losses of nutrients in natural ecosystems, comprising the soil–plants and animals–atmosphere, are roughly in balance so that continued growth depends on the cycling of nutrients between the *biomass*, *organic* and *inorganic* stores. Removals from agricultural systems, especially of crop products, generally exceed natural inputs unless these are augmented by fertilizers and organic manures.

Nutrients in the biomass store pass to the organic store by excretion from or on the death of living organisms. The greater part of the organic and inorganic stores is in the soil. Additions to the inorganic store occur through mineralization of organic residues, weathering, rainfall and dust fallout. That part of the inorganic store on which plants feed is called the *available* or *labile* pool, consisting of ions in solution and adsorbed by clays, sesquioxides and organic matter, as distinct from nutrients held in insoluble precipitates or strongly adsorbed complexes, which are considered *non-labile*.

The ability of a minority of plants and micro-organisms to reduce N_2 gas to NH_3 within the cell (*nitrogen fixation*) provides a unique input of N to the biomass and thence organic stores. But substantial losses of N can occur following *nitrification* due to the leaching of NO_3^- in the soil solution and NO_3^- reduction (*denitrification*) to N_2O and N_2 under anoxic conditions.

Phosphate is rapidly immobilized by soil micro-organisms, by adsorption and by precipitation of insoluble Ca, Fe or Al phosphates. Many plants have evolved fungus–root associations or *mycorrhizas* which enhance P absorption from deficient soils. In contrast to sulphate, which is steadily leached except from acid sesquioxidic soils, P losses by leaching are negligible because the P concentration in solution is very low.

Calcium, Mg and K are held mainly as exchangeable cations, the supply of which buffers the soil solution against depletion. Cation losses are accelerated when anions such as NO_3^-, SO_4^- and HCO_3^- are plentiful in the percolating water, and under acid conditions when H^+ and Al^{3+} ions occupy exchange sites. Small amounts of the micro–nutrients Fe, Mn, Cu, Zn and Co are also held as exchangeable cations, and as organo-metal complexes in solution. Molybdenum and boron occur as anions in solution.

REFERENCES

COLE D.W., GESSEL S.P. & DICE S.F. (1967) Distribution and cycling of nitrogen, phosphorus, potassium and calcium in a second-growth Douglas fir ecosystem, in *Primary Productivity and Mineral Cycling in Natural Ecosystems* (Ed. H.E. Young). Ecological Society of America, New York.

LIKENS G.E., BORNMANN F.H., JOHNSON N.M. & PIERCE R.S. (1967) The calcium, magnesium, potassium and sodium budgets for a small forested ecosystem. *Ecology* **48**, 772–785.

NYE P.H. & GREENLAND D.J. (1960) The soil under shifting cultivation. *Commonwealth Bureau of Soils, Technical Communication No.* **51**, Harpenden, England.

QUISPEL A. (1974) (Ed.) The biology of nitrogen fixation. *Frontiers of Biology* **33**, North Holland, Amsterdam.

FURTHER READING

HAYMAN D.S. (1975) Phosphorus cycling by soil micro-organisms and plant roots, in *Soil Microbiology* (Ed. N. Walker). Butterworths, London.

HODGSON J.F. (1963) Chemistry of the micronutrient elements in soils. *Advances in Agronomy* **15**, 119–159.

MITCHELL R.L. (1970) Trace elements in soils and factors that affect their availability. *Geological Society of America, Special paper* **140**, 9–16.

NICHOLSON T.H. (1967) Vesicular-arbuscular mycorrhiza—a universal plant symbiosis. *Science Progress, Oxford* **55**, 561–581.

NUTMAN P.S. (1965) Symbiotic nitrogen fixation, in *Soil Nitrogen* (Eds. W.V. Bartholomew and F.E. Clark). *Agronomy No.* **10**, American Society of Agronomy, Madison.

VINCENT J.M. (1965) Environmental factors in the fixation of nitrogen by the legume, in *Soil Nitrogen* (Eds. W.V. Bartholomew and F.E. Clark). *Agronomy No.* **10**, American Society of Agronomy, Madison.

Chapter 11
Maintenance of Soil Productivity

11.1 TRADITIONAL METHODS

Bush fallowing

The practice of *shifting cultivation* is widespread in tropical regions, accounting for some 30 per cent of the world's exploitable soil resources. The cycle begins with the clearing and burning of the natural vegetation, which in the case of the rainforest releases a great store of available nutrients for the growth of crops such as maize, cowpeas, yams and groundnuts. Although these are usually sown with a minimum of soil disturbance, nutrient losses by leaching and soil erosion are high and the encroachment of weeds rapid, so that after one or two crops the land is abandoned to the regenerating forest and the peasant farmer moves to a new site, where the cycle is repeated.

Regrowth of the native vegetation, especially the forest or *bush fallow*, is vital to the stability of such an agricultural system because the soil's fertility is restored by the deep-rooting trees or perennial grasses, drawing up nutrients from deep in the subsoil and returning large quantities of litter to the soil surface. There is also an input of N fixed symbiotically in the soil of humid forest regions. Experience in the savanna regions suggests that a cycle of 2–4 years cropping and 6–12 years fallow can maintain fertility in the long term, but in the wet forest zones a 1–2 year cropping period and 10–20 years fallow are preferred.

Rotational cropping

In the more closely settled temperate regions, competition for farming land forced the adoption of intensive cropping systems much sooner than in the Tropics. One of the earliest was the simple 3 year rotation of autumn cereal—spring cereal—fallow, practised in England since Celtic times. This wholly arable system was largely displaced in the 18th century by rotations, such as the Norfolk four course, in which grass–clover pastures and root crops were alternated with cereals,

thus permitting the close integration of livestock and arable farming. Added advantages were the nitrogen fixed by the legume and the abundant residues left in the soil by the grass.

However, depressed cereal prices in the 1880s and a growing belief in the value of grass as a 'soil conditioner' led to the temporary pasture or *ley* being extended for 3 or more years: gradually the arable–ley rotation evolved into the practice of *mixed farming*. The essence of this system is the rearing of livestock on farm-produced grass and cereals, in the course of which much of the nutrient gathered from the soil is returned in organic manures.

Organic manures

Farmyard manure (FYM), deep litter (from stall-fed animals), poultry (broiler) manure, sludge and compost are examples of organic manures (Table 11.1). Their

Table 11.1. Range and median element contents of organic manures.

	Dry matter	N	P	K
		(% fresh weight)		
FYM	11–92 (23)	0.2–3.5 (0.6)	0.04–1.3 (0.13)	0.08–3.6 (0.6)
Poultry manure	5–96 (29)	0.1–6.8 (1.7)	0.04–3.4 (0.6)	0.03–3.6 (0.6)
Liquid cattle manure and slurry	1–60 (4)	0.005–4.8 (0.3)	0.0002–1.1 (0.04)	0.0008–3.5 (0.25)
Pig slurry	1–69 (4)	0.01–4.8 (0.4)	0.004–2.1 (0.09)	0.017–2.7 (0.17)
Compost	24–95 (28)	0.6–1.1 (0.9)	0.18–0.35 (0.22)	0.17–0.75 (0.33)
Sewage sludge	5	0.25	0.16	0.01

(after MAFF Bulletin 210)

144

beneficial effects lie as much in the improvement of structure accruing from large additions of organic matter, as in the contribution made to the soil's nutrient supply. Organic manures also encourage flourishing populations of small animals, especially earthworms, as well as micro-organisms that produce essential growth factors (hormones and vitamins) which may stimulate plant growth.

(a) *FYM* consists of cattle dung and urine mixed with straw. It is very variable in composition and although it is low in macronutrients, at normal application rates of 40–50 t ha^{-1}, the quantities of micronutrients supplied are roughly equivalent to those removed in a succession of 4 to 5 crops (Table 11.2).

Table 11.2. Micronutrients in FYM compared with those removed in crops.

	Mn	Zn	Cu	Mo
	(kg ha^{-1})			
FYM (45 t ha^{-1})	3.36	1.12	0.56	0.01
Total of 4 arable crops in succession	2.50	1.80	0.30	0.01

(after Cooke, 1975)

(b) *Slurry* is a suspension of dung in the urine and washing water coming from animal houses and milking parlours. Compositions of some representative slurries, liquid manures and poultry are given in Table 11.1.

(c) *Sludge* consists of the solids separated from liquid raw sewage. *Digested* sludge has been fermented anaerobically to eliminate offensive odours and lower the count of pathogens and may safely be applied to the land. Nevertheless, sustained heavy applications of sludge, especially that from industrial sewage, may raise Zn, Cu, Ni and Cd to levels that cause plant and animal nutritional disorders. The average N, P and K contents of digested sludge are given in Table 11.1.

(d) *Compost* is made by accelerating the rate of humification of plant and animal residues in well aerated, moist heaps. Ideally, a compost of low C:N ratio and acceptable nutrient content can be made in 6 to 8 weeks (Table 11.1).

Green manure

The practice of ploughing in a quick-growing leafy crop before maturity is called *green–manuring*. A major benefit is the avoidance of NO_3-N leaching in susceptible soils, for the nitrate is taken up by the green manure crop and released to a subsequent cash crop as the low C:N ratio residues decompose. If the green manure crop is a legume there can be an additional boost to the soil N supply from symbiotically fixed nitrogen.

The long-term effect on soil organic matter is minimal, as the succulent residues are rapidly decomposed and contribute little to the soil humus. Similarly, effects on structure through the stimulation of soil microbial activity are ephemeral.

Recent developments

Shifting cultivation and the bush fallow are becoming increasingly untenable in developing countries because of the pressure for land and food exerted by burgeoning populations. Rarely can land and scarce soil water reserves be diverted to grow green manure crops on a large scale, and in those areas where freedom from debilitating diseases allows cattle to be kept, the dung has traditionally been burnt as domestic fuel.

With the intensification of agriculture in temperate regions, especially after the Second World War, livestock and arable farming have become increasingly divorced, so that animal manures are not plentiful in the areas where they are most needed. Inevitably, there has been a change-over to cereal monoculture—'the one course rotation'—and increasing reliance on chemical fertilizers in place of traditional manuring and fallowing. Production has exceeded overall demand, generally with increased profitability, although this is an achievement not without cost in terms of more acid soils (section 11.3), deterioration in soil structure (section 11.4) and accelerated erosion (section 11.5). Before discussing such problems, it is appropriate to review the means of assessing soil fertility.

11.2 PRODUCTIVITY AND SOIL FERTILITY

Soil nutrient supplying power

Soil, climate, pests, disease, genetic potential of the crop and man's management are the main factors governing *productivity*, as measured by the *yield* of crop or animal produce per hectare. Where the variables other than soil remain reasonably constant, one expects a direct dependence of yield on soil fertility.

It is often the case that the ability of the soil to supply one nutrient, especially N, P or K, has an overriding effect on fertility: the yield then increases linearly with the supply of that nutrient, until the supply of another nutrient becomes limiting. This concept, embodied in Liebig's 'law of the minimum', is illustrated in Figure 11.1. *Soil testing* has as its objectives firstly

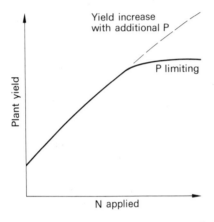

Figure 11.1. Yield response to N in the presence and absence of additional P.

to identify which nutrients are deficient, and secondly to quantify the relationship between yield and the limiting nutrient supply, that is,

Yield = f (soil N supply)
P, K and other nutrients non-limiting   (11.1)

Soil testing

DEFICIENCY DIAGNOSIS
Various signs of disorder (chlorotic leaves, stunted growth or withered petioles) are indicative of the deficiency of one or more essential elements. Visual symptoms appear, however, only after the plant has suffered a check in growth due to 'hidden hunger' for the deficient element. Growth and the concentration of the element in the plant's tissues are closely linked, as shown in Figure 11.2, so that *plant analysis* or *tissue testing* is useful in diagnosing deficiency.

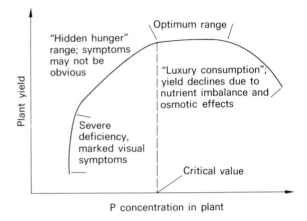

Figure 11.2. The relationship between yield and plant tissue concentration of phosphorus.

Alternatively, the soil may be extracted with chemical solvents, or shaken in suspension with anion or cation exchange resins, with the aim of removing only that part of the total element content that is available to the plant. Radioactive isotopes such as ^{32}P may also be used to label the available fraction, or *labile pool* (section 12.3), the amount of which is calculated from the quantity of tracer added and the specific activity of the soil solution measured when dilution of the isotope throughout the labile pool is complete.

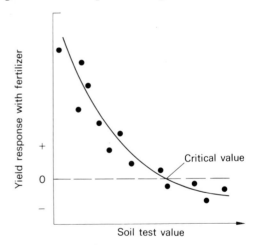

(Black circles represent individual soils)

Figure 11.3. Yield response to fertilizer in relation to soil test value (points represent individual soils tested).

Labile (isotopically-exchangeable) P

$$= {}^{32}\text{P soln} \times \frac{{}^{31}\text{P soln}}{{}^{32}\text{P soln}} \quad (11.2)$$

Whatever the method, however, soil and tissue tests must be calibrated against crop yields, usually on a variety of soils with different rates of fertilizer, whence a *critical value* for the test is established. This is the value above which the crop is not likely to respond to fertilizing with the element in question (Figure 11.3).

QUANTITATIVE ASSESSMENT OF FERTILIZER NEEDS

If the soil test predicts a response to fertilizing (with P for example), one then needs to know how much fertilizer to apply. This is most accurately estimated by means of a *fertilizer rate trial*, in which the yields at different rates of fertilizer (amounts per hectare) are measured. The fertilizer rate affording maximum yield can be determined; but more importantly, the rate giving the *maximum profit* per hectare can also be calculated. In Figure 11.4, for example, when the value

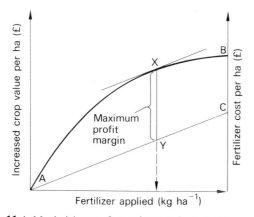

Figure 11.4. Maximizing profit per hectare from fertilizer investment.

of the extra yield due to fertilizer (curve AB) is compared with the fertilizer cost line (AC), the vertical XY indicates the fertilizer rate giving the biggest value–cost differential per hectare. If the crop's value, or the fertilizer's cost changes, XY slides to the right or left to a new point maximum of profit.

11.3 SOIL ACIDITY AND LIMING

Origin of acidity

A soil becomes acid as Ca^{2+}, Mg^{2+}, K^+ and Na^+ ions are leached from the profile faster than they are released by mineral weathering, and H^+ and Al^{3+} ions become predominant on the exchange surfaces (section 7.2). The intensity of the acidity, measured by the soil pH, is the resultant of an interaction of mineral type, climate and vegetation. The main causes of acidification may be summarized as follows:

(a) Microbial oxidation of organic matter, producing humic residues with acidic carboxyl and phenolic groups, and leading to a higher CO_2 partial pressure in the soil air which displaces the carbonic acid equilibria to the right:

$$CO_2 + H_2O \rightleftharpoons H^+ + HCO_3^- \rightleftharpoons H^+ + CO_3^= \quad (11.3)$$

(b) Nitrification of NH_4^+ ions, producing H^+ ions, and NO_3^- which is susceptible to leaching (equation 8.10).

(c) Increased anion concentrations in the soil solution (HCO_3^-, NO_3^- and $SO_4^=$ from dissolved SO_3 in rainwater) which must be balanced by equivalent concentrations of cations as they are leached through the soil.

(d) In soils formed on marine muds, or coal-bearing sedimentary rocks, the oxidation of iron pyrites (formed under intense reducing conditions in the original deposit—section 8.4) gives rise to *acid sulphate, soils* according to the reactions:

$$2FeS_2 + 7O_2 + 2H_2O \rightleftharpoons 4SO_4^= + 4H^+ + 2Fe^{2+} \quad (11.4)$$

$$2Fe^{2+} + \tfrac{1}{2}O_2 + 5H_2O \rightleftharpoons 2Fe(OH)_3 + 4H^+ \quad (11.5)$$

Under natural conditions, plants have evolved and adapted to the prevailing soil pH. The *calcifuges* are tolerant of acidity compared to the *calcicoles* which are intolerant of acidity, or 'lime-loving'. Cultivated plants show similar diversity in their response to acidity as the selection of crop and pasture species listed in Table 11.3 shows. Within the legume family, the tropical species are more tolerant of acidity than the medics and clovers of temperate regions.

Liming

The problems associated with soil acidity—slow turnover of organic matter, poor nodulation of some

Table 11.3. Sensitivity of cultivated plants to soil acidity.

Species	pH below which growth is restricted
Rye	4.9
Potato	4.9
Ryegrass	5.1
Wheat	5.4
Maize	5.5
Oats	5.5
White clover	5.6
Barley	5.9
Sugar beet	5.9
Trefoil	6.1
Lucerne	6.1

(after MAFF Bulletin 209)

legumes, Ca and Mo deficiencies, Al and Mn toxicities —can be remedied by liming. Ground limestone, chalk, marl and basic slag are used as liming materials, the active constituent being primarily $CaCO_3$, with some burnt lime (CaO) and hydrated lime $(Ca(OH)_2)$. On the addition of lime to acid soil, the pH slowly rises according to the reaction:

$$CaCO_3 + H_2O \rightleftharpoons Ca^{2+} + HCO_3^- + OH^- \qquad (11.6)$$

When sufficient lime is added to leave some residual carbonate in the soil, the pH attains a maximum c. 8.3 if the CO_2 partial pressure is as low as that of the atmosphere. Any increase in CO_2 causes more carbonate to dissolve, according to the reaction:

$$CaCO_3 + CO_2 + H_2O \rightleftharpoons Ca(HCO_3)_2 \qquad (11.7)$$

and a lower pH is attained, as seen from the equation:

$$pH = K - \tfrac{1}{2}\log (P_{CO_2}) - \tfrac{1}{2}\log (Ca) \qquad (11.8)$$

where P_{CO2} = partial pressure of CO_2, (Ca) = activity of Ca^{2+} ions and $K = 4.85$ if the carbonate has the solubility of pure calcite.

LIME REQUIREMENT

The *lime requirement* is dependent on the pH buffering capacity of the soil (section 7.2), and is expressed as the quantity of $CaCO_3$ (t ha^{-1} to a depth of 15 cm) required to raise the soil pH to a desired value. The standard method of measurement is described by Hooper (1973).

A pH of 6.5 is recommended for temperate crops, but liming of grassland to pH 6 is preferred, since grasses are more tolerant of acidity and the possibility of inducing deficiencies of Mn, Cu and Zn is minimized. Tropical species are more acid-tolerant and furthermore, liming to pH 6–6.5 may reduce P availability in soils high in exchangeable Al^{3+} (section 10.3) and destabilize the structure of kaolinitic clay soils (section 7.4).

Finely ground $CaCO_3$ made to adhere to legume seeds—a process called *lime pelleting*—greatly improves nodulation of sensitive species in acid soils, especially when an effective strain of *Rhizobium* bacteria is included in the coating.

11.4 THE IMPORTANCE OF SOIL STRUCTURE

Structure dependent properties

Management imposes constraints on productivity, irrespective of the soil's nutrient supply. Good soil management should aim to create optimum physical conditions for plant growth, as manifest by
(1) adequate aeration for roots and micro-organisms;
(2) adequate available water;
(3) ease of root penetration, permitting thorough exploitation of the soil for water and nutrients;
(4) rapid and uniform seed germination;
(5) resistance of the soil to slaking, surface-sealing and accelerated erosion by wind and water.

Two most useful indices of structure are *bulk density*, which is inversely related to total porosity (section 4.5), and *aggregate stability*, which is measured either by sieving the soil under water (the wet sieving technique) or by immersing aggregates in solutions of varying ionic strength (the Emerson dispersion test). These methods are described in texts by Baver (1972) and Loveday (1974). Bulk density measurements are more relevant to conditions (1), (2) and (3) above, whereas aggregate stability is more relevant to (4) and (5).

Organic matter, texture and land use modify these properties, and it is found, for example, that bulk density increases as organic matter content decreases, and is highest in the sandy textural classes. Aggregate stability, on the other hand, decreases with declining organic matter and is least for soils in the silty and fine sandy loam classes.

Aeration and water supply depend on the soil's pore size distribution, the key indices being the *air capacity* $C_a(50)$ (section 4.5) and the *AWC* (section 6.5). Using these two properties, classes of soil droughtiness at one extreme and susceptibility to waterlogging at the other, may be separated, as illustrated for surface soils in England and Wales which regularly experience an *SMD* > 100 mm (Figure 11.5). Soils with < 10 per cent

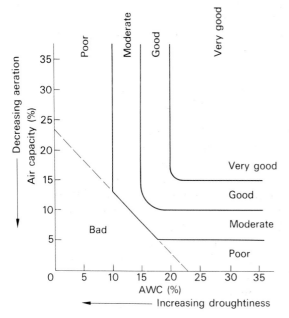

Figure 11.5. Classification of the structural quality of topsoils (after Hall *et al.*, 1977).

AWC are droughty, and those with < 5 per cent air space at field capacity are likely to be anaerobic. The limiting *AWC* value is lowered slightly and the $C_a(50)$ value raised to 10 per cent for soils of wetter regions. Irrespective of the individual values of $C_a(50)$ and *AWC*, a *storage pore space* value (the sum of *AWC* and $C_a(50)$) less than 23 per cent is undesirable, hence the truncation of the curve separating the 'poor' and 'moderate' classes in Figure 11.5.

The effect of land use

GRASS VS ARABLE

Soil physical conditions are usually best under permanent grass (or forest) and deteriorate at a rate dependent

on climate, soil texture and management as the soil is cultivated.

Disruption of peds exposes previously inaccessible organic matter to attack by micro-organisms; populations of structure-stabilizing fungi decline and earthworm numbers are markedly depressed. Removal of the protection of a permanent canopy of vegetation exposes the soil to the direct impact of rain, and the loss of surface litter reduces the detention time of water before surface runoff and concomitant erosion begins. Shearing forces generated by tractor wheels and ploughs, especially when the soil is plastic, lead to the sundering of intraped bonds, the re-orientation of clay domains and the formation of dense layers of low vertical permeability called *plough pans*.

Measurements of water-stable aggregates show that the deterioration in structure is most rapid in the first year out of grass (Figure 11.6). Total porosity and *AWC* also decline as cultivation is prolonged. These trends are reversed, however, by the introduction of a ley into the cropping cycle, the restorative effect being dependent on the interaction between soil and climate, the species in the ley and its duration. In temperate regions, it is true to say 'the longer the ley, the better', but in tropical soils near maximum benefit from a ley accrues within two years and the effect disappears rapidly once the soil is cultivated again, presumably

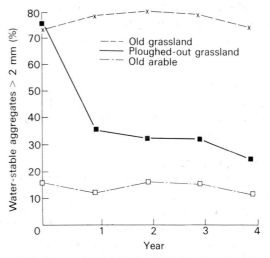

Figure 11.6. Aggregation of the fine earth fraction of a soil under different land use (after Low, 1972).

due to the high rate of organic decay. In the latter case, a two year ley: two year cropping cycle is recommended.

MODIFIED METHODS OF TILLAGE

The conventional method of cultivation or tillage in Western agriculture involves the thorough disturbance of the top 20–25 cm of soil, usually with the inversion of the furrow slice, as in mouldboard ploughing (Figure 11.7a). Residues of the previous crop and volunteer weeds are buried, and after repeated passes with discs and harrows the soil is reduced to a fine tilth suitable for seeding.

Because this process is very demanding of time, labour and energy, it may impose restraints on the farm cropping programme as well as contributing to structural deterioration. New tillage methods have therefore

Figure 11.7. (a) Inverted furrow slices following mouldboard ploughing (courtesy of D.E. Patterson). **(b)** Soil after direct drilling (courtesy of R.Q. Cannell).

(a)

(b)

been devised which preserve the natural soil structure, yet provide suitable conditions for seed germination, and in many cases confer the added benefit of leaving residues on the surface to protect the soil from erosion and reduce the evaporation of water. There practices range from *zero tillage* or *direct drilling* (Figure 11.7b) where the only mechanical operation is seed sowing, to *reduced* or *minimum* tillage which involves fewer and shallower passes with tines or discs than does conventional ploughing. The term 'reduced tillage' may also refer to the practices of *strip cropping* and *stubble mulching* which are discussed in section 11.5.

DIRECT DRILLING

The replacement of ploughing by direct drilling on arable land initiates gradual changes which are generally beneficial to soil structure. Organic matter increases in the top 5 cm and the water stability of the surface peds is enhanced. On some soils, bulk density is increased down to the old plough depth so that total porosity decreases, mainly at the expense of the macropores. Nevertheless, water infiltration and percolation do not necessarily suffer because earthworms multiply two or three fold, especially the deep burrowing *Lumbricus terrestris*, and the predominantly vertical orientation and continuity of their holes makes for rapid water flow. Furthermore, the higher bulk density improves the 'trafficability', i.e. load bearing capacity, of the soil under wet conditions.

Changes in other properties of the no-till soil are less favourable. Immobile elements such as P and K become concentrated in the 0–5 cm layer and may become less available to the plant as the soil dries out. Total N follows the trend in organic C, but nitrate levels are often lower possibly because of localized denitrification and greater downward leaching. Consistent with greater NO_3^- leaching are the lower exchangeable Ca and pH values in the 0–5 cm depth of some direct drilled soils, which may therefore require additional lime. Other possible disadvantages are:
(a) slow warming of the wet soil in spring which retards crop growth;
(b) the production of volatile fatty acids around fermenting crop residues in the drill silt, which can inhibit seed germination (section 8.3);

(c) the difficulty (and cost) of weed control that is reliant solely on herbicides.

To sum up, it is fair to say that the choice between direct drilling and traditional cultivation is delicately balanced in the intensive agriculture of humid temperate regions. However, under more extreme climates where steep slopes and soil structural instability heighten the erosion hazard, direct drilling generally produces higher yields and certainly minimizes soil degradation, as indicated in Table 11.4.

Table 11.4. Average soil losses under different soil management in Illinois, U.S.A. (annual precipitation *c.* 1300 mm).

	Average annual soil loss (t ha^{-1})	
	5% slope	9% slope
Conventional tillage, maize and wheat	7.6	21.5
No-till, maize and wheat	0.9	1.3
No-till, continuous maize	0.6	0.9

(after Gard and McKibben, 1973)

11.5 SOIL EROSION

The term *erosion* describes the transport of soil constituents by natural forces, primarily water and wind. As indicated in section 5.2, past phases of geologic erosion led to the formation of sedimentary deposits which are the parent materials of many present-day soils; but in the short term erosion due to man's activities, especially agriculture, is of greater concern. In the U.S.A., for example, one-fifth of the 175 million ha of cropland loses more than 20 t ha^{-1} of soil annually, one-half loses between 7.5 and 20 t and the remainder less than 7.5 t. Erosion is very damaging to soil fertility because it is mainly the nutrient-rich surface soil which is removed, and of that, predominantly the fine and light fractons—clay and organic matter, leaving behind the inert sand and gravel.

Erosion by water

A global study of stream sediment loads suggests the general relation between erosion by water and rainfall shown by the solid line in Figure 11.8. In arid areas, nearly all the rain falling is absorbed by the parched

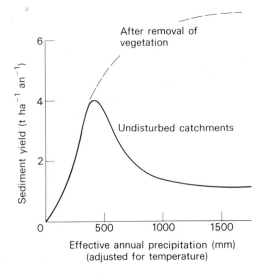

Figure 11.8. Sediment loads in rivers in relation to precipit-ation (adjusted for differences in temperature) (after Langbein and Schumm, 1958).

(a)

(b)

Figure 11.9. (a) Soil surface just before raindrop impact. **(b)** Soil surface immediately after raindrop impact showing splash erosion (courtesy of D. Payne).

soil and scanty vegetation, leaving little to run off, while in humid climates abundant vegetation protects the soil all the year. Most vulnerable are the regions of inter-mediate rainfall where the incidence of rain is seasonal and the vegetative cover therefore less effective, espec-ially at the end of each dry season. Nevertheless, the erosive power of high rainfall is attested by the broken curve of Figure 11.8 which applies to humid regions when the natural vegetation is removed.

Soil loss by erosion is dependent on (a) the potential of rain to erode—the rainfall *erosivity*, and (b) the susceptibility of soil to erosion—the soil *erodibility*.

RAINFALL EROSIVITY

To erode soil, work must be done to disrupt aggregates and move soil particles: the energy is provided by the kinetic energy (*K.E.*) of falling raindrops and flowing water. However, the energy available from falling rain-drops, which causes *splash erosion* (Figure 11.9) is at least 200 times greater than that of surface runoff, which causes *wash, rill* or *gully erosion* (Figure 11.10).

Raindrop *K.E.* increases with the mass of the raindrop and the square of its *terminal velocity*. Both character-istics depend on the rate of rainfall or *intensity*, ex-pressed in mm hr^{-1} (that is, m$^3 \times 10^3$ of rain per m^2 of

surface perpendicular to the rain's direction per hour). The frequency of large raindrops, up to 5 mm diameter, increases as the intensity rises to approximately 100 mm hr^{-1}. Terminal velocity also increases with drop size so that the overall effect is one of the *K.E.* per mm of rain rising sharply up to an intensity $\sim$100 mm hr^{-1} (Figure 11.11).

Another important characteristic of rainfall is the frequency of high intensity storms, for which there is a striking difference between tropical and temperate regions. Some 95 per cent of temperate rain but only 60 per cent of tropical rain falls at intensities <25 mm hr $^{-1}$, which are considered non-erosive. Further, the maximum intensity of tropical rainfall may exceed 150 mm hr $^{-1}$ which is twice that of temperate rainfall. The high average K.E. and generally greater amounts of tropical rainfall make erosion by water *potentially* a very serious problem in these regions, whereas it is of little importance in humid temperate lands.

SOIL ERODIBILITY
Erodibility is defined by the equation:

$$\frac{A}{R} = E.I. \tag{11.9}$$

where A = annual soil loss in t ha $^{-1}$;
$\quad$ R = dimensionless rainfall erosivity index;
$\quad$ $E.I.$ = *soil erodibility index*, the value of which depends on soil type (K), as modified by relief features (LS), erosion control measures (P) and crop management (C). Written in the form:

$$A = R\ K\ LS\ P\ C \tag{11.10}$$

the equation is called the *Universal Soil Loss Equation* (*USLE*) from which A can be calculated from known values of R, K, LS, P and C. Alternatively, by setting an upper limit for the loss as T (usually between 2.5 and 12.5 t ha $^{-1}$), the cropping and soil conservation practices (*CP*) necessary to keep $A \leqslant T$ can be determined from the relationship:

$$CP = \frac{T}{RKLS} \tag{11.11}$$

The use of *USLE* in modern soil conservation practice is discussed at length by Hudson (1971); it is sufficient here to examine briefly the interrelation between erodibility and the variables K, LS, P and C.

(a) *Soil factor K*. K values depend primarily on texture, organic matter, structural stability and permeability of the soil. Texture is not amenable to change, but the effect of leys or permanent pasture on the other variables affecting K has already been noted (section 11.4). Temporary amelioration of the surface structure of arable soils can be achieved by applying soil conditioners such as PVA in aqueous solution (Figure 11.12) although the cost is considerable.

Figure 11.10. Gully erosion caused by surface runoff on bare soil (courtesy of A.J. Low).

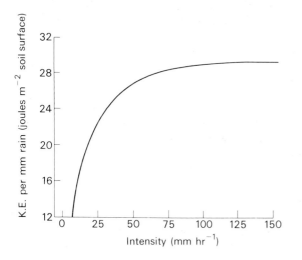

Figure 11.11. Kinetic energy per mm of rain in relation to rainfall intensity (after Hudson, 1971).

ment possible is to plough straight up and down the slope ($P=1$). Practices which can reduce P by 50 to 75 per cent range from ploughing across the slope, or along contours, to elaborate arrays of terraces or banks designed to hold surplus water as long as possible and to divert exceptional flows down grassed waterways

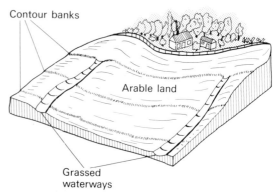

Figure 11.13. Perspective of crop land with contour banks and waterways.

(Figure 11.13). Contour banks are most effective on slopes of 2 to 8 per cent. Much earth movement and grading may be required, so that the final design is a compromise between the cost of the work, including the loss of cropping flexibility incurred, and the value of the land and its produce.

(d) *Crop management C.* Manipulation of this factor achieves the most significant reductions in erosion loss, as evidenced by Table 11.4. The C value may vary from 0.001 for well kept woodland to 1.0 for continuous bare fallow. Practices aimed at reducing C must focus on
(1) not leaving the soil bare when the rainfall erosivity is high, and
(2) growing high density, healthy crops so that the soil surface is protected and the binding of the soil by the roots is strong.

Leys, permanent pasture, minimum tillage and direct drilling all have merit for erosion control. To these must be added *stubble mulching*, whereby the residues of a previous crop are chopped up and scattered over the soil surface, and *strip cropping* where strips of arable land and pasture alternate down a slope, thereby reducing the length of slope most vulnerable to erosion.

Figure 11.12. Stabilization of surface soil structure—the darker plots have been treated with PVA solution (courtesy of J.M. Oades).

(b) *Slope factor LS.* Both steepness (S) and length (L) of slope affect erodibility—the former because of the greater potential energy of soil and water high on a slope, which is convertible to kinetic energy; the latter because the longer the slope, the greater the surface runoff and its downhill velocity. When soil conservation measures are introduced, such as contour terracing or strip cropping, the distance L of vulnerable soil between structures is adjusted according to the slope S so that a composite factor LS is used in the *USLE.*

(c) *Conservation practice P.* The worst soil manage-

(a)

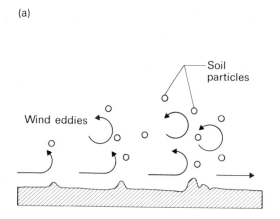

(b)

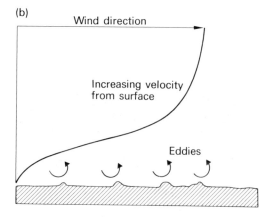

Figure 11.14. (a) Turbulent wind eddies over a roughened soil surface (after Hudson, 1971). (b) Vertical profile of wind velocity over a soil surface (after Hudson, 1971).

Erosion by wind

Soil movement by wind is prominent in arid areas and on coastlines where the wind is strong and consistent and the vegetation adversely affected by salinity. By analogy with water erosion, the severity of wind erosion depends on the *potential* of the wind to erode and the *susceptibility* of the soil to erosion. Although not as widespread as water erosion, the effects of wind erosion are often more dramatic in that the quantities of soil moved per hour across the edge of one hectare can be of the same order as that moved by water erosion in one year.

WIND ACTION

Small protruberances in an otherwise smooth surface create turbulent eddies in the wind which lift up light, loose particles (Figure 11.14a). The air-borne particle gains a forward velocity, and because of the increase in wind velocity away from the soil surface (Figure 11.14b) the higher the jump, the greater the particle's momentum. Apart from wind velocity, the height of the jump will depend on the particle's size, expressed by its

$$effective\ diameter = \left(\frac{\text{particle } B.D. \times \text{diameter}}{\text{specific gravity}}\right).$$

Three consequences follow from the uplift and gain in momentum of the erodible particles:
(a) If the effective diameter is <0.06 mm, the particle

stays suspended for a long time, resulting in an *aerial dispersion* which if sufficiently concentrated becomes a dust storm. In times past, especially during the Pleistocene glaciations, the larger wind-dispersed particles (0.02–0.06 mm) settled to form the extensive *loess* deposits described in section 5.2. Smaller particles may be blown across oceans and continents before being brought down in rain.

(b) For particles and aggregates between 0.06 and 0.2 mm effective diameter, the jump is followed by a long flat trajectory to earth. On striking the surface, the particle either bounces off or comes to rest, transmitting its *K.E.* to other particles which are knocked upwards (Figure 11.15). This self-sustaining process is called *saltation*.

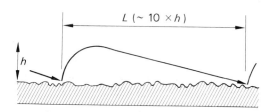

Figure 11.15. Jump and glide path of saltating soil particles (after Hudson, 1971).

(c) On landing, the saltating particle may strike larger, heavier particles which do not rise up but merely roll or *creep* along the surface. Particles and aggregates of effective diameter between 0.2 and 1 mm move in this way.

Movement by saltation accounts for more than 50 per cent of wind erosion, with 3–40 per cent of movement occurring in suspension and 5–25 per cent by surface creep. Little can be done to lessen the erosive potential of wind other than by reducing its velocity by means of *wind breaks*, and by shortening its *fetch*, that is the distance of its unhindered travel over erodible land.

SOIL FACTORS

Moist soil supporting healthy vegetation is the most effective deterrent to wind erosion. Cultivated soil is vulnerable when the surface is dry and the structure weak, especially if most of the aggregates are < 1 mm. Often coarse and fine textured soils are more erodible than loams which have the right proportions of clay, silt and sand to form stable aggregates of a non-erodible size.

Surface roughness also affects erodibility, for although slight irregularities initiate saltation as the wind speed rises, very rough or cloddy surfaces protect against erosion by reducing the average wind velocity at ground level and trapping any saltating particles as they land. *Vegetation*, even if dead, has a similar effect which is directly proportional to its density and height above the soil. It follows that stubble mulching, pioneered on the wind-swept High Plains of the U.S.A., and direct drilling beneath old crop residues, are valuable in controlling wind erosion.

11.6 SUMMARY

Clearing a soil of natural vegetation interrupts the cycling of nutrients and may lead to a decline in soil fertility. Traditional methods of farming in the Tropics (*shifting cultivation* and *bush fallowing*) and temperate regions (*rotational cropping* and *mixed farming* with the use of organic and green manures) generally maintain fertility, but have gradually been displaced by intensive monotypic systems, such as continuous cereal cropping, in response to the pressure for land, the cost and scarcity of labour and the rising demand for food.

The nutrient budget can be balanced by the use of inorganic fertilizers, the need for which can be quickly identified by *soil testing* or *plant analysis* for key elements such as P and K, and the required amounts assessed from *fertilizer rate trials*. Regular use of NH_4-N fertilizer increases soil acidity and consequent problems of Al and Mn toxicity, Ca and Mo deficiency, which are easily corrected by *liming*.

Deterioration of soil structure is a more insidious problem which is only revealed through inadequate soil aeration, reduced *AWC*, impedance of root penetration and surface sealing which inhibits seedling emergence and predisposes the soil to erosion by water and wind. A grass ley following arable cropping is very effective in restoring structure, the improvement being more rapid in the Tropics, where a 2 year ley: 2 year cropping cycle is acceptable, than in temperate regions where the ley should last for 3 or more years. *Reduced tillage* and *direct drilling* help to preserve a good structure and are effective in reducing erosion losses.

Erosion is very damaging to soil fertility because it is mainly the nutrient-rich surface soil that is removed. Water erosion occurs in almost any environment, if the soil is bare and the structure unstable, the magnitude of the loss depending on the rainfall *erosivity* and soil *erodibility*. This relationship is quantified by assigning values to the variables in the Universal Soil Loss Equation:

$$A \quad = \quad R \quad \times (K, LS, P, C)$$
(soil loss) (Erosivity) (Erodibility)

Conservation measures aim at reducing the slope steepness S and length L, the soil management factor P (through contour banks, terraces, strip cropping) and the crop management factor C (by direct drilling and stubble mulching).

Wind erosion is confined to arid regions or to coastal areas where wind velocities are consistently high. The susceptibility of soil to wind erosion depends on the effective diameter of aggregates (< 0.6 mm are most vulnerable) and the degree of protection afforded the surface by vegetation.

REFERENCES

BAVER L.D., GARDNER W.H. & GARDNER W.R. (1972) *Soil Physics*. 4th Ed., Wiley, New York.

HOOPER L.J. (1973) Fertilizer recommendations. *MAFF Bulletin* **209**, HMSO, London.

HUDSON N.W. (1971) *Soil Conservation*. Batsford, London.

LOVEDAY J. (1974) (Ed.) Methods for analysis of irrigated soils. *Commonwealth Bureau of Soils, Technical Communication No.* **54**.

FURTHER READING

BAEUMER K. & BAKERMANS W.A.P. (1973) Zero tillage. *Advances in Agronomy* **25**, 77–123.

HALL D.G.M., REEVE M.J., THOMASSEN A.J. & WRIGHTS V.F. (1977) Water retention, porosity and density of field soils. *Soil Survey of England and Wales, Technical Monograph No.* **9**.

LOW A.J. (1972) The effect of cultivation on the structure and other physical characteristics of grassland and arable soils (1945–1970). *Journal of Soil Science* **23**, 363–380.

RUSSELL R.S. (1977) *Plant Root Systems*. McGraw-Hill, London.

STRUTT N. (1970) Modern farming and the soil. *Report of the Agricultural Advisory Council on Soil Structure and Soil Fertility*. HMSO, London.

TISDALE S.L. & NELSON W.L. (1975) *Soil Fertility and Fertilizers*. 3rd Ed. Macmillan, London.

Chapter 12
Agricultural Chemicals and the Soil

12.1 WHAT ARE AGRICULTURAL CHEMICALS?

Broadly, agricultural chemicals may be subdivided into two categories:
(a) *fertilizers*, briefly introduced in section 11.1, which are used to correct nutrient deficiencies and imbalances in plants and animals; and
(b) *pesticides* (sometimes called plant-protection chemicals), which are used to protect crops from pests and disease or to eliminate competition from aggressive weeds.

For all fertilizers, excluding foliar sprays, and the bulk of the pesticides, reactions in the soil are the key to determining their efficacy and their fate in the environment. It is the aim of this chapter to discuss the nature and soil reactivity of these chemicals, starting with the N, P, K fertilizers.

12.2 NITROGEN FERTILIZERS

Forms of N fertilizer

SOLUBLE FORMS
The chemical compositions of the common *straight* N fertilizers are given in Table 12.1: these consist of a single active constituent. *Mixed* or *compound* fertilizers containing various proportions of N, P and K salts (which interact to some extent in the mixture) are also widely used, either as solids and more recently as liquids, specifically *nitrogen solutions* such as NH_4NO_3 and urea dissolved in water.

Highly soluble fertilizers can pose handling and storage problems because of their hygroscopicity, a difficulty largely overcome by *granulation*, which is intended to reduce the contact area between individual particles and also the area available for water absorption. Ideally, a granulated (or granular) fertilizer has uniform granules, roughly spherical in shape and *c.* 2.5–3 mm in diameter: the ultimate in granulation is the spherical *prills* of NH_4NO_3 and urea. Granulated straight fertilizers may be mixed to give *bulk-blended* fertilizers which are popular in the U.S.A.

Table 12.1. Forms of soluble N fertilizer.

Compound	Formula	N content
Solids		(%)
Sodium nitrate	$NaNO_3$	16
Calcium nitrate	$Ca(NO_3)_2$	17
Ammonium sulphate	$(NH_4)_2SO_4$	21
Calcium cyanamide	$CaCN_2$	21
Ammonium nitrate	NH_4NO_3	35
Urea	$(NH_2)_2CO$	46
Monoammonium phosphate	$NH_4H_2PO_4$	11–12
Diammonium phosphate	$(NH_4)_2HPO_4$	18–21
Liquids		
Aqua ammonia	NH_3 in water	25–29
Anhydrous ammonia	liquefied NH_3 gas	82

The choice of the most suitable N fertilizer depends on a balance of factors—the cost per kg of N (including transport and application cost), the effects on plant growth (both beneficial and detrimental) and the magnitude of N loss through leaching and denitrification. The advantage of a very high N content per unit weight afforded by anhydrous NH_3, for example, is partly offset by higher costs of application, since it must be kept under pressure and injected some 15 cm below the soil surface. Other effects are discussed below.

SLOW-RELEASE FERTILIZERS
To obviate problems arising from the extreme solubility of many N fertilizers, and the susceptibility of NO_3–N to leaching, various sparingly soluble N compounds including natural organic materials have been developed (Table 12.2). These are usually inferior, per kg of N, to

the soluble forms, but they are especially useful for the establishment of vegetation on difficult sites, such as

Table 12.2. Forms of slow-release N fertilizer.

Fertilizer	Composition	N content (%)
Shoddy	wool waste	2–15
Dried blood	by-products of	~13
Hoof and horn meal	meat processing	7–16
Ureaform	ureaformaldehyde polymers	21–38
IBDU	isobutylidene diurea	32
SCU	sulphur-coated urea	37–40

reclaimed mine workings, where a slow but assured release of N over several seasons is required. An alternative solution to the leaching problem is to use NH_4–N fertilizers only, together with the specific inhibitor of nitrification 'N-serve' (2-chloro-6 trichloromethyl pyridine), at the rate of 1–2 per cent of the N applied.

Reactions in the soil

SCORCH AND RETARDED GERMINATION

There is a possibility of retarded germination, injury to young roots and leaf scorch* due to the concentrated salt solution around dissolving fertilizer granules, particularly if they are combine-drilled into the soil with the seed (Figure 12.1). Generally fertilizers containing NO_3–N are more harmful than NH_4–N fertilizers; calcium nitrate is satisfactory below 75 kg N ha^{-1} while $(NH_4)_2SO_4$ is safe up to 125 kg N ha^{-1}. Urea is exceptional in being injurious at rates as low as 35 kg N ha^{-1}. The best method of applying urea is to place it in a band, slightly below and to one side of the seed (Figure 12.1) where damage due to high salt concentration and losses by volatilization of NH_3 are minimized. Mixed fertilizers containing N are less injurious to plants than comparable rates of straight fertilizers.

* Chlorosis followed by leaf death, usually extending from the tip backwards.

Figure 12.1. Ways of applying fertilizer to soil.

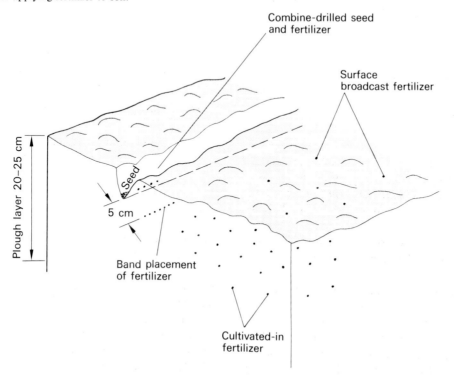

PH EFFECTS

Ammonia solutions, which contain NH_4OH, create an alkaline reaction in the soil, as does urea which is rapidly hydrolysed to ammonium carbonate $(NH_4)_2CO_3$ (see equation 10.4). As the pH rises above 7, NH_4^+ ions become increasingly unstable, decomposing to H^+ and NH_3 gas which diffuses out of the soil rapidly.

A pH rise to 7.7 and above in the vicinity of ammonium or urea fertilizer will inhibit *Nitrobacter* organisms; *Nitrosomonas* is not so sensitive and functions well up to pH 9 so that NO_2^- accumulates in the soil. Although oxidation of NH_4^+ to NO_2^- gradually reduces the pH, the high level of NO_2^- may in turn continue to inhibit *Nitrobacter*.

NITRIFICATION, DENITRIFICATION AND LEACHING

Irrespective of its initial form, nearly all the fertilizer N applied to a cultivated soil finally appears as NO_3^-, the implications of which for denitrification and leaching loss have been discussed in section 10.2. *Direct* leaching, that is the removal of N fertilizer in surface runoff, is only likely to occur on arable soils when heavy rainfall closely follows fertilizer application, a coincidence sometimes observed with spring cereals in the U.K. (Figure 12.2). More insidious, however, is the slow percolation of NO_3^- into groundwaters, such as in the Chalk aquifers in England, where NO_3^- levels have risen steadily over the past decade. Underground supplies make up 40 per cent of all potable water in England and Wales so that the maintenance of good water quality is vital to public health. A nitrate concentration < 11.3 ppm N is recommended by the World Health Organization for European waters, with 11.3–22.6 ppm being acceptable and > 22.6 ppm unacceptable. At the higher levels, babies become susceptible to *methaemoglobinaemia* or 'blue-baby disease', a condition induced by reduction of NO_3^- to NO_2^- in the stomach and the absorption of excess NO_2^- into the blood.

Nitrification also leads to accelerated cation leaching (section 10.4). Theoretically, for every kg N in the ammonium form that is oxidized to nitrate *and leached*, some 7 kg $CaCO_3$ equivalent is lost from the soil. It is for this reason that mixtures of NH_4NO_3 with $CaCO_3$ ('Nitro-Chalk') have proved popular with farmers in the

U.K. In practice, losses of Ca are much lower because some of the nitrate is taken up by plants and some is denitrified.

Efficiency of utilization

RATE AND METHOD OF APPLICATION

The recovery of fertilizer N in the crop or animal product

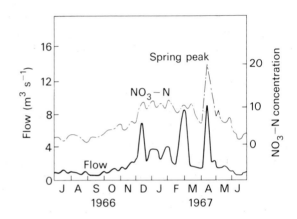

Figure 12.2. Nitrate levels in a river draining fertile arable land (after Tomlinson, 1972).

depends on the climate, the kind of crop and on the rate, method and timing of the fertilizer application. Potentially the best recovery is achieved when temperature and other factors allow the crop to be grown during the wetter part of the year, as can be done in much of the Tropics. For a given climate, N recoveries are usually lowest for plantation crops and short season arable crops, and highest for permanent grass. Bananas in the Ivory Coast, for example, recovered only 14 per cent of the 430 kg N ha^{-1} applied, but grass recovered 53 per cent at rates up to 700 kg ha^{-1}. In Britain, where leaching losses during the wet winter months are appreciable, N recovery ranges from about 50 per cent on leys receiving an average of 140 kg N ha^{-1} annually, to 70 per cent on permanent grass receiving around 70 kg N ha^{-1} an^{-1}.

Timing of N applications is crucial for best utilization by the crop. Spring applications in Britain give better results than autumn, especially in areas with > 675 mm annual precipitation (and > 125 mm of excess winter rainfall*). Little nitrate remains in the soil in spring when the excess winter rainfall exceeds 250 mm. During the grass growing season, N is best applied as a *split dressing*, say 2 or 3 times during the season, or after each cut of hay or silage that is made. In this way, the supply of N is better matched to the growth cycle of the grass and losses by leaching are minimized.

Similarly, in the southern maize belt of the U.S.A. it has been found that delaying N application until 3–4 weeks after germination avoids leaching losses from late spring rains, which can be heavy, and provides N at the stage of growth when the plant's demand is highest.

RESIDUAL EFFECTS

A residual effect occurs when the fertilizer applied to one crop benefits the growth of a succeeding crop. For N fertilizer, residual effects depend on the rate of N applied, the crop, the rain falling between cropping periods and the soil type. Based on extensive data from catchment studies, it is possible to construct for Britain a schedule of probable leaching losses as demonstrated in Table 12.3. As excess winter rainfall increases, one

Table 12.3. Estimated N losses by leaching in relation to excess winter rainfall in Britain.

Mean NO_3-N concentration in drainage water	Quantity of N leached by an excess winter rainfall (mm an^{-1}) of:				
	100	200	300	400	500
(ppm)	(kg ha^{-1})				
5	5	10	15	20	25
10	10	20	30	40	50
15	15	30	45	60	75

moves diagonally across the table from left to right because NO_3–N concentrations invariably rise as the runoff rate increases. Thus, for an excess winter rainfall of 250–300 mm and allowing for crop uptake, there will be little residual N from a fertilizer application of 100 kg N ha^{-1} to a preceding crop. The situation can be worse than Table 12.3 suggests because appreciable water flow and NO_3^- leaching occurs through well structured soils at water contents below field capacity.

* Defined as the excess of precipitation over evaporation when the soil is at field capacity during the winter months.

Residual effects have been detected when deep-rooting cereals such as winter wheat follow root crops to which heavy N dressings have been applied (Table 12.4a). In

Table 12.4. Residual effects of N-fertilized crops and legumes.

(a)	Wheat yield following potatoes given		
	0	and	185 kg N ha^{-1}
		(t grain ha^{-1})	
	2.84		3.91
(b)	Wheat yield following N-fertilized ryegrass		clover ley
		(t grain ha^{-1})	
	4.83		6.12

(after Cooke, 1969)

other cases, such as cereal following cereal, part of any residual effect is undoubtedly due to the larger residue of N-rich plant material in the soil which decomposes to release mineral N. When the preceding crop is a legume, the stimulatory effect of the residues is generally greater than that of even a heavily fertilized grass (Table 12.4b).

12.3 PHOSPHATE FERTILIZERS

Forms of phosphate fertilizer

Phosphate fertilizers may be subdivided into the *orthophosphates*, the condensed orthophosphates or *polyphosphates*, and the insoluble *mineral* and *organic phosphates* (Table 12.5). The natural rock phosphates, consisting of the minerals *apatite* and *fluorapatite* with calcite and other impurities, are the raw material from which other P fertilizers are made by processes that render part or all of the P soluble in water.

ORTHO-P FERTILIZERS

The active constituents are *monocalcium phosphate* (MCP), formula $Ca(H_2PO_4)_2.H_2O$, and *phosphoric acid*. The former compound is the basis of the *superphosphates* formed by the dissolution of rock phosphate in H_2SO_4 or H_3PO_4; for example

$$Ca_{10}(PO_4)_6F_2 + 7\ H_2SO_4 + 3\ H_2O \rightarrow 3\ Ca(H_2PO_4)_2.H_2O$$
$$+ 7\ CaSO_4 + 2\ HF \quad (12.1)$$
(single superphosphate)

Table 12.5. Forms of phosphate fertilizer.

Fertilizer	Composition	P content (%)
Ortho-P		
Phosphoric acid	H_3PO_4	23
Normal or single superphosphate	$Ca(H_2PO_4)_2$; $CaSO_4$	8–10
Concentrated or triple superphosphate	$Ca(H_2PO_4)_2$	19–21
Nitric phosphate	$NH_4H_2PO_4$; $CaHPO_4$ and NH_4NO_3	6–11
Monoammonium phosphate	$NH_4H_2PO_4$	21–26
Diammonium phosphate	$(NH_4)_2HPO_4$	20–23
Poly-P		
Superphosphoric acid	$H_4P_2O_7$ and higher M.W. polymers	> 33
Ammonium polyphosphate	$(NH_4)_4P_2O_7$ and higher M.W. polymers	26–27
Insoluble phosphates		
Rock phosphate ore	$Ca_{10}(PO_4)_6(F,OH)_2$; $CaCO_3$ and sesquioxide impurities	6–18
Basic slag	Basic Ca, Mg phosphates, Fe_2O_3 and SiO_2	3–10
Organic phosphates	Bone meal: guano	5–8

$$Ca_{10}(PO_4)_6F_2 + 14\ H_3PO_4 + 10\ H_2O$$
$$\rightarrow 10\ Ca\ (H_2PO_4)_2.H_2O + 2\ HF \quad (12.2)$$
$$\text{(triple superphosphate)}$$

Single superphosphate has the advantage of containing 14 per cent S which is of additional value on sulphur-deficient soils.

Phosphoric acid is made by treating rock phosphate ore with sulphuric acid or by roasting the ore in an electric furnace to form P_2O_5, which is then dissolved in water to form H_3PO_4, according to the reaction:

$$P_2O_5 + 3H_2O \rightarrow 2\ H_3PO_4 \quad (12.3)$$

Phosphoric acid normally contains 52–54% P_2O_5 (23–24% P). Partial neutralization of the acid with NH_3 gives the ammonium phosphates with $NH_3:H_3PO_4$ mole ratios ranging from 1 (*monoammonium phosphate*, MAP) to 2 (*diammonium phosphate*, DAP). These compounds, which combine a dual nutrient supply with the advantages of high solubility, good handling properties and low production cost, have rapidly increased in popularity since their introduction in the early 1960s.

Nitric phosphates are made by dissolving rock phosphate in concentrated HNO_3 (usually with some H_2SO_4 or H_3PO_4) and neutralizing the mixture with NH_3, according to the reaction:

$$Ca_{10}(PO_4)_6F_2 + 20\ HNO_3 + 4\ H_3PO_4 + 21\ NH_3$$
$$\rightarrow 9\ CaHPO_4 + CaF_2 + NH_4H_2PO_4 + 20\ NH_4NO_3 \quad (12.4)$$

This fertilizer was introduced in the 1950s when a world shortage of S for H_2SO_4 manufacture was thought imminent, but now remains popular only in continental Europe.

The relationship between these ortho-P fertilizers is summarized in Figure 12.3. They supply the bulk of the world's P fertilizer.

POLY-P FERTILIZERS

When P_2O_5 is combined with less water than is used to form orthophosphoric acid (equation 12.3), *superphosphoric acid* is formed consisting of a mixture of orthophosphoric acid and higher molecular weight polymers, as for example:

$$3P_2O_5 + 5H_2O \rightarrow 2H_5P_3O_{10} \quad (12.5)$$
$$\text{(tripolyphosphoric acid)}$$

Ammoniation of superphosphoric acid produces very soluble fertilizers of high N and P analysis—the *ammonium polyphosphates*. Much less soluble polyphosphate fertilizers, such as the calcium and potassium metaphosphates, have also been produced but are not widely used.

INSOLUBLE PHOSPHATES

Rock phosphates are associated with some igneous and sedimentary rocks. In sedimentary deposits, the crystals of apatite and fluorapatite are cemented together by calcium carbonate to form granules. Granule size and the magnitude of the clay, sesquioxide and silica impurities determine the solubility of the ore in water and acids. *Beneficiation*—the removal of much of the impurity by flotation and washing in water—is the first

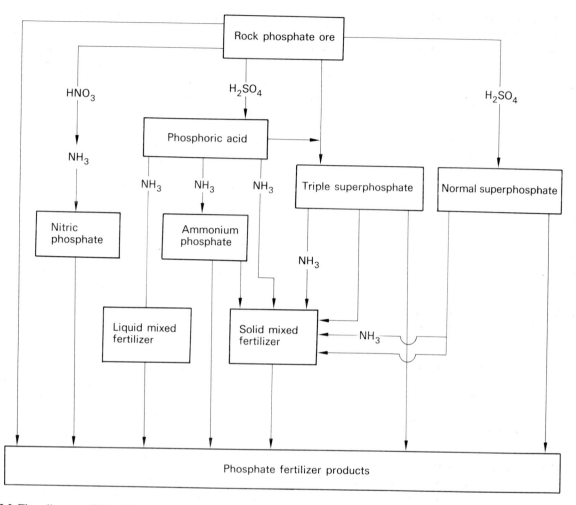

Figure 12.3. Flow diagram of P fertilizer manufacture (after Slack, 1967).

step in raising the P content of the ore (to 13–18 per cent). Finely ground beneficiated rock phosphate (>90 per cent passing through a 100 mesh sieve) is used as a fertilizer.

Ground rock phosphate (GRP) costs less per kg of P than soluble P fertilizers, but it is only successful on acid soils for slow-growing grass or tree crops, for which repeated dressings of soluble P fertilizer are uneconomic. Other rock phosphate fertilizers are made by heating finely ground ore to 1200°C and above—a process called *calcination*, or by heating the ore with soda ash and silica (the product is called Rhenania

phosphate in Germany). The roasting ignites organic impurities and converts fluorapatite to the more soluble *β-tricalcium phosphate*, $Ca_3(PO_4)_2$.

Sometimes GRP is mixed with *basic slag*, a by-product of steel manufacture from phosphatic iron ores. Basic slag must also be finely ground, and it is best used on acid soils where its liming effect (equivalent to approximately two-thirds its weight of ground limestone) and its micronutrient impurities are beneficial, especially for legumes.

The efficacy of these insoluble fertilizers during the first season after their application is roughly related to their *citrate solubility*, which is measured in 2 per cent citric acid in the U.K., 1 M ammonium citrate at pH 7 in the U.S.A., or ammonium citrate at pH 9 on the Continent. The availability to plants of several insoluble phosphates compared to superphosphate is illustrated in Figure 12.4.

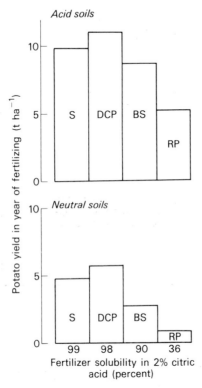

[S = superphosphate; DCP = dicalcium phosphate; BS = basic slag; RP = rock phosphate]

Figure 12.4. Potato yields in relation to the citrate solubility of P fertilizers (after Mattingly, 1963).

Reactions in the soil

DISSOLUTION OF WATER SOLUBLE FERTILIZERS
When a granule of MCP, the active constituent of superphosphate, is placed in dry soil it takes up water by vapour diffusion, or by osmosis in a moist soil until a saturated salt solution forms which flows outwards from the granule. This solution is very acid (pH ~ 1.5) and concentrated in P (*c*. 4 M) and Ca (*c*. 1.4 M). It reacts

with the soil minerals, dissolving large amounts of Ca, Al, Fe and Mn, some of which are subsequently precipitated as new compounds—the *soil-fertilizer reaction products*, as the pH slowly rises and solubility products are exceeded. An MCP granule of 5 mm diameter dissolves in 24–36 hours to leave a residue of DCPD containing about 20 per cent of the original P. The sequence of events is summarized in Figure 12.5.

The rapidity of this reaction means that the plant feeds not so much on the fertilizer itself in the soil, but on the fertilizer reaction products which are metastable and slowly revert to thermodynamically more stable (but less soluble) products. Depending on whether the soil is calcareous or rich in sesquioxides, for example, and on the presence of other salts such as NH_4Cl, $(NH_4)_2 SO_4$ or KCl within the granule, intermediate products ranging from potassium and ammonium *taranakites* ($H_6K_3Al_5(PO_4)_8.18H_2O$ and $H_6(NH_4)_3 Al_5(PO_4)_8.18H_2O$), complex calcium–aluminium and calcium–iron phosphates, to *octacalcium phosphate* (OCP) are formed. The Al compounds slowly hydrolyse to amorphous $AlPO_4$ which eventually reverts to *variscite*, the Fe compounds hydrolyse to amorphous $FePO_4$ which reverts to *strengite*, while DCPD and OCP revert to *hydroxyapatite* (HA), with the result that, taking account of the interaction of fertilizer type, time and soil, a wide spectrum of P availability confronts the plant, as illustrated in Figure 12.6.

POWDERED VS GRANULAR FORMS
Insoluble fertilizers such as the basic calcium phosphates are most effective when used in a finely divided form. Hygroscopic absorption of water is not the problem that it is with soluble fertilizers, but because fine powders are difficult to apply, a form of weak aggregation or 'mini-granulating' of the powdered fertilizer is sometimes employed. By contrast, granulation of soluble P fertilizers has the advantage, in addition to ease of handling, of confining the reaction between soil and fertilizer to small volumes around each granule. The rate of phosphate fixation in the soil is therefore diminished.

RESIDUAL EFFECTS
Only some 10 to 20 per cent of fertilizer P may be

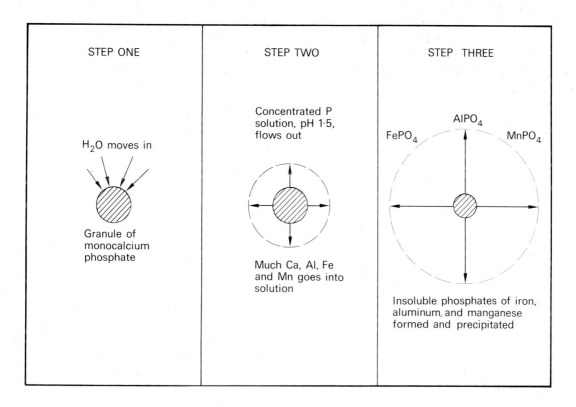

Figure 12.5. Successive stages in the dissolution of an MCP granule in soil (after Tisdale and Nelson, 1975).

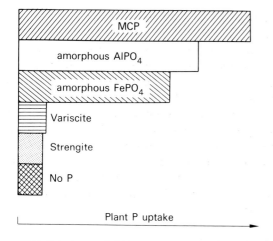

Figure 12.6. Relative availability of different P sources to maize (after Huffman, 1962).

absorbed by plants during the first year: the remainder is nearly all retained as fertilizer reaction products which become less soluble with time. A rough guideline is that two-thirds of the water and citrate-soluble P remains after one crop, one-third after two crops, one-sixth after three crops and none after four crops.

Larsen (1971) has suggested that a more precise way of assessing the residual effect is to measure the rate at which the labile pool of soil phosphate (L value) declines after the addition of fertilizer, and to calculate the half-life for the 'decay' of the fertilizer's value to the crop. Half-lives in neutral and alkaline soils vary from one to six years. Larsen also pointed out that insoluble P fertilizers, in contrast to soluble ones, required a certain time to reach peak effectiveness, as well as having a half-life for immobilization (Figure 12.7).

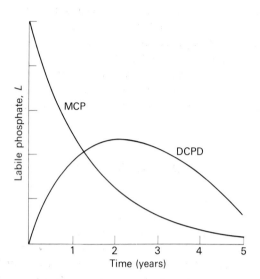

Figure 12.7. Changes in soil labile phosphate after the addition of P fertilizer (after Larsen, 1971).

Eutrophication

CAUSE AND EFFECTS

Eutrophication is a *natural* process whereby surface waters are gradually enriched with organic and inorganic nutrients carried in runoff from the surrounding catchment. Natural eutrophication is evidenced by the formation of fen peats, in both temperate and tropical climates, where base-rich waters collect in drainage depressions. Nevertheless, eutrophication can become a problem when it is *accelerated* by increased loads of sediment and dissolved nutrients in the streams draining urban and rural areas: the higher levels of P, N, Si and organic matter promote the growth of aquatic plants, from minute phytoplankton to rooted higher plants.

Surges of phytoplankton growth are called *algal blooms*. These detract from the amenity appeal of the water and pose problems for water treatment works, because of the fouling of primary filter beds and tainting of the water by essential oils released by the algae. Proliferation of aquatic weeds makes a stream more sluggish, with consequently greater sediment deposition and eventual blocking of the channel. Dead vegetation sinks to the bottom where a thick layer of decomposing organic sludge accumulates, consuming large quantities of dissolved oxygen and creating anoxic conditions which may result in fish death.

Experience indicates that accelerated eutrophication and the appearance of algal blooms are most likely to occur
(a) in standing bodies of water;
(b) during the high light, high temperature conditions of summer;
(c) when the P concentration of the water exceeds *c.* 0.05 ppm and the ratio of dissolved N:P is ~20 or less.

SOURCES OF NUTRIENT INPUT

Dissolved and particulate nutrient loads arise from *point* and *non-point* (diffuse) sources. Examples of point sources are effluents from factories, sewage works and intensive livestock units such as large cattle feedlots and dairies. Non-point sources comprise the drainage from the remaining agricultural and non-agricultural land—grassland, field and forest, which by its very nature is difficult to monitor accurately.

In the U.K., the bulk of the N-enrichment of surface waters comes from agricultural land where each year drainage from one hectare may carry between 10 and 100 kg N, depending on the land use and fertilizers applied. Conversely, between two-thirds and four-fifths of the phosphate in surface waters appears to come from point sources, especially intensive animal farms, which release much organic P, and sewage works which handle the domestic and industrial detergent load. Orthophosphate ions are normally immobile in soil and drainage from agricultural land usually contributes < 1 kg P ha^{-1} an^{-1}. Exceptionally losses may rise into the 1–10 kg P ha^{-1} an^{-1} range when P-rich surface soil is eroded, when organic manure is spread copiously on soil that is either frozen or very wet, or when P fertilizer is liberally applied to very acid peaty soils.

12.4 POTASSIUM FERTILIZERS

Forms of K fertilizer

The major K compounds listed in Table 12.6 present a wide range of K content and water solubility. Potassium is mined as *muriate of potash* KCl, which is high in K and very soluble; but chloride ions impair the quality of

Table 12.6. Forms of potassium fertilizer.

Fertilizer	Composition	K content
		(%)
Potassium chloride	KCl	52
Potassium sulphate	K_2SO_4	44
Potassium calcium pyrophosphate	$K_2CaP_2O_7$	21–25
Potassium metaphosphate	KPO_3	28–33
Kainit	Mineral potassium salt with $< 3.6\%$ Mg	10–25

crops such as tobacco and potatoes, so that K_2SO_4 may be preferred as a soluble K fertilizer despite its higher cost per kg of K. The solubility of potassium metaphosphate (KMP) and potassium calcium pyrophosphate (KCP) varies according to the impurities present and the granule size, granules < 0.4 mm in diameter being as soluble as KCl. *Kainit* is an insoluble K fertilizer.

K balance in cropping
Since the potassium immediately available to plants is held as exchangeable cations, which are not readily leached nor rendered unexchangeable, it is feasible to attempt to balance K removed by crops and in animal products, with K applied in fertilizers and released from non-exchangeable sources in the soil.

K RELEASE
Slow release of non-exchangeable K depends on the content of potash feldspars and micaceous clay minerals. Soils with such reserves, even though low in exchangeable K, may supply between 20 and 80 kg ha^{-1} an^{-1} for many years, especially to grasses which are efficient in taking up K at low concentrations; but sandy soils lacking micaceous clays are soon exhausted of K by continuous cropping. On many soils where potassium is adequate initially, regular N and P fertilizing may so stimulate growth that K deficiency symptoms appear, more often on pasture land where fodder conservation (hay or silage) is practised.

K UPTAKE BY CROPS
Recovery of K fertilizer during one season varies from 25–80 per cent, being highest for grassland, which may remove up to 450 kg K ha^{-1} an^{-1} when highly produc-

tive. Clovers, cereals and root crops require higher soluble K levels in soil than grasses, so that the ratio of N:K fertilizer applied should be 2:3 compared to 1:1 for temperate grasses and 2:1 for tropical grasses. If too much K is used on grazed grassland, luxury uptake of K may induce Mg deficiency in the herbage which predisposes to *grass tetany* or *hypomagnesaemia* in the stock.

12.5 PESTICIDES IN THE SOIL

Environmental impact
The term 'pesticide' embraces the many natural or synthetic chemical compounds which have biocidal or biostatic effects: they can be more specifically designated *insecticides, miticides, nematicides, fungicides, herbicides, rodenticides* or *molluscicides* according to the group of organisms they are intended to control. Of these, the herbicides, insecticides and fungicides account for the bulk of agricultural usage.

The properties that commend a chemical as an effective pesticide, namely
(a) a *broad spectrum* of activity so that it controls more than one pest, and
(b) sufficient *persistence* at the site of application so that frequent retreatment is unnecessary,
are generally inimicable to the ecological balance of the whole environment. Broad spectrum pesticides kill harmless and beneficial organisms as well as the target organisms, and the greater their persistence, the longer the duration of their deleterious effects. The ideal pesticide is one which controls only the target organism and persists only long enough to complete this purpose before degrading into harmless products.

Before World War II, the main pesticides in use were inorganic compounds of arsenic and copper, oils derived from petroleum and coal-tar (creosote), together with a few plant-derived compounds such as pyrethrum and nicotine. The natural organic compounds were either rapidly decomposed in the environment or if persistent, of very low mammalian toxicity, so that only the residues of Cu from the fungicidal Bordeaux mixture ($CuSO_4$ dissolved in lime water) or of As from Ca and Pb arsenates persisted as potential environmental pollutants. But events during and after World War II, starting with the discovery of the insecticidal properties of DDT in 1939, resulted in a burgeoning of the quantity and variety of synthetic chemicals available for pest control (Table 12.7). Many

Table 12.7. Growth in world production of pesticides.

Year	1945	1955	1965	1975
Quantity of pesticides (kilotonnes)	100	400	1000	1800

(after Green, Hartley & West, 1977)

of the new chemicals, epitomized by the *organochlorine insecticides*, were very stable compounds and those of the cyclodiene group (dieldrin, aldrin and endrin) were also highly toxic to mammals; the organochlorines

suffer the added disadvantage of being lipid-soluble and generally accumulating in the fatty tissues of animals, particularly predators high up in natural food chains. Indiscriminate application of the new chemicals led to their residues and breakdown products becoming widely disseminated, with many instances of their detrimental effect on beneficial insects and plants, domestic animals and wildlife. The strong public outcry against such malpractice has resulted in governments exercising stricter controls over the manufacture and release of new chemicals. Nowadays, procedures to evaluate *environmental impact* and *pesticidal efficacy* of the kind outlined in Figure 12.8 must be followed before a new compound is approved for commercial use. The environmental impact depends on the quantity of chemical used and its toxicity to non-target organisms, as well as on its persistence in the environment. The

Figure 12.8. Pesticide testing procedures in the U.K. (after Glasser, 1976).

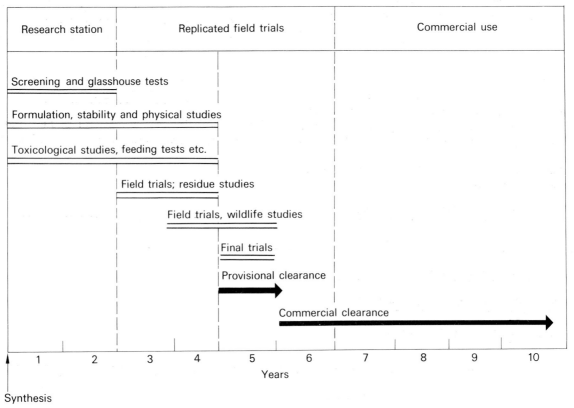

factors governing pesticide stability in the soil and mechanisms of loss to surface waters and air are discussed below.

Pesticide persistence

The main pathways of pesticide loss from soil are:
(a) Volatilization, especially of surface-applied compounds
(b) Chemical and biological decomposition
(c) Absorption by plants and animals
(d) Transport on eroded soil particles
(e) In solution, by leaching or in surface runoff

The relative importance of each pathway depends very much on the *properties of the chemical*—its water-solubility, volatility, affinity for organic or mineral surfaces; its *formulation* and *mode of application*—whether presented as a spray, dust or in granules, applied to crop surfaces or cultivated into the soil; the *environmental conditions* (temperature and rainfall), the *soil type* and *cropping system*. Maximum persistence times of the major pesticides, measured as the time for 90 per cent or more of the chemical to disappear from the site of application, are shown in Figure 12.9.

ORGANOCHLORINE INSECTICIDES

Of the four important compounds listed in Table 12.8, DDT, dieldrin and lindane are broad spectrum chemicals of great stability in the soil: their persistence ranges from 4–6 years for lindane to 8–10 years for DDT and

Table 12.8. Description and key properties of four important organochlorine insecticides.

Common name	Chemical name	Solubility in water	Vapour pressure at 30°C
		(ppm)	(mm Hg)
DDT	1,1-bis(4-chlorophenyl)-2,2,2-trichloroethane	0.001–0.04	7×10^{-7}
Dieldrin (HEOD)	1,2,3,4,10,10-hexa-chlorocyclopentadiene	0.1–0.25	1×10^{-5}
Lindane (γ-BHC)	γ-1,2,3,4,5,6-hexa-chlorocyclohexane	7.3–10.0	13×10^{-5}
Toxaphene	chlorinated camphene containing *c.* 68% chlorine	0.4	no data

(after Guenzi and others, 1974)

dieldrin. Toxaphene is now the most widely used of the group because of its low mammalian toxicity and non-persistence in animal fats.

Being unionized and non-polar they are weakly adsorbed by soil organic matter. The affinity of adsorption may be predicted from the partition coefficient K_d of the compound between water and a non-miscible organic solvent, that is

$$K_d = \frac{[\text{Pesticide concentration}]\ \text{octanol}}{[\text{Pesticide concentration}]\ \text{water}} \quad (12.6)$$

Nevertheless, leaching of these compounds is negligible because of their very low water solubility. They are also

Figure 12.9. Maximum persistence of pesticides in soil under a mild climate (after Stewart *et al.*, 1975).

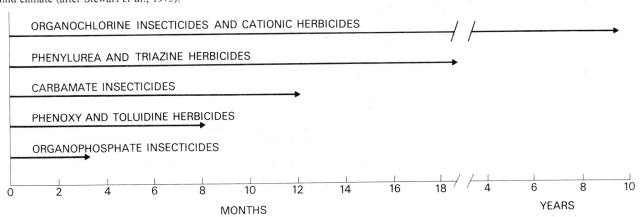

extremely resistant to chemical and microbial de-composition.

One of the major pathways of organochlorine loss from soil is by *volatilization*. Eighty per cent or more of a 5 kg ha^{-1} application may be lost to the air by spray drift and volatilization from spray drops during application; and volatilization from the residue in moist soil can be at the rate of 10 kg ha^{-1} an^{-1} *initially*, for a relatively volatile compound such as lindane. The rate of volatilization is much reduced when the pesticide is incorporated into the soil, and especially when there is less than the equivalent of a monolayer of water molecules adhering to clay surfaces (the soil is air-dry). At very low soil water contents, the organochlorines are strongly adsorbed due to van der Waals' forces and there is a large reduction in their vapour pressure in the soil air (Figure 12.10).

BIPYRIDYL HERBICIDES

The bipyridyl herbicides, *diquat* and *paraquat*, are of similar persistence in soil to the organochlorines but

for different reasons. They are slightly soluble cationic compounds which are strongly adsorbed on clay surfaces by electrostatic attraction.

$$[CH_3-N \quad \bigcirc-\bigcirc \quad N-CH_3]^{2+} + 2Cl^- + Ca^{2+}clay \rightleftharpoons$$
$$[CH_3-N \quad \bigcirc-\bigcirc \quad N-CH_3]^{2+}clay + CaCl_2 \qquad (12.7)$$
$$\text{paraquat}$$

Adsorbed, the molecule is biologically inert; but adsorption also induces a shift in the wavelength of maximum light absorption by paraquat into the visible range, so that it is decomposed photochemically if exposed to sunlight.

TRIAZINE AND PHENYLUREA HERBICIDES

These two groups contain some of the most widely used modern herbicides. The triazines, of which *simazine* is

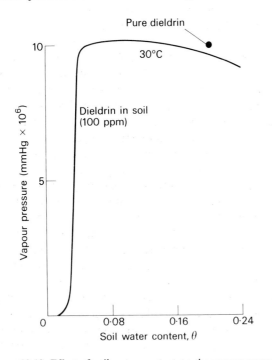

typical are weak bases of relatively low solubility (5 ppm for *simazine* to 677 ppm for *prometone*) which become protonated at low pH, according to the reaction

$$B + H_3O^+ \rightleftharpoons [BH]^+ + H_2O \qquad (12.8)$$

Adsorption of the protonated molecule by clays and organic matter accounts for the persistence of these chemicals in the soil. The major degradative pathway of the chloro-*s*-triazines (including *atrazine* and *propazine*) is by substitution of OH for the Cl atom on the ring and microbial de-alkylation of the side chains.

Typical of the phenylureas is *monuron*

$$Cl \bigcirc NHCON(CH_3)_2$$

These nonionic compounds are weakly adsorbed by soil organic matter and range in solubility from 42 ppm for *diuron* to *c*. 3300 ppm for *fenuron*. The more soluble compounds are therefore leached from soils; microbial deactivation also occurs by de-alkylation, e.g.

Figure 12.10. Effect of soil water content on the vapour pressure of dieldrin (after Spencer, 1970).

fenuron

and cleavage of the amide bond

aniline

CARBAMATE AND ORGANOPHOSPHATE INSECTICIDES

These two groups are treated together because they inhibit several ester-splitting enzymes in animals, most importantly cholinesterase, which is essential for muscle co-ordination. The *organophosphates* are related to the 'nerve gases' and have the basic structure of an esterified PO_4 group where R_1 and R_2 are alkyl groups and X is

one of a range of substituted organic (often heterocyclic) groups. They are more volatile and water-soluble than the organochlorines, and despite their high mammalian toxicities, are favoured over the latter because of their rapid degradation in soil. The most widely used are *parathion*, *methylparathion*, *diazinon* and *demeton*.

The *carbamates* show a very wide spectrum of biocidal activity; the methylcarbamates of which *carbaryl*

is an example, are primarily insecticides which have found favour because of their specificity for pest organisms and their short persistence in the soil.

TOLUIDINE AND PHENOXYALKANOIC ACID HERBICIDES

The toluidines are important pre-emergent herbicides, controlling both grass and broad-leaf weeds. The basic structure is that of a 2,6-dinitroaniline:

Variation in the substituents X and Y alter the compound's selectivity and activity, X being CF_3 and Y being H in the most widely used compound, *trifluralin*. The toluidines are insoluble in water (< 1 ppm) and are adsorbed primarily by organic matter. Volatilization and microbial breakdown provide the main pathways of loss.

The other important group, the *phenoxy acids*, are selective for broad-leaf weeds in cereal crops and grassland. The most common representatives are:

2,4-D   2,4-dichlorophenoxyacetic acid
2,4,5-T   2,4,5-trichlorophenoxyacetic acid
MCPA   (4-chloro-o-tolyl)oxyacetic acid

These compounds are negatively charged in soil and therefore are not appreciably adsorbed except by sesquioxides, or by weak hydrogen-bonding to organic matter. In spite of this, and their moderate water solubility (400–4500 ppm), these chemicals show little leaching loss because they are absorbed by plants and are rapidly decomposed by micro-organisms.

Biodegradation of these herbicides illustrates a phenomenon common to all pesticide molecules when they are added to a soil not previously exposed to the compound in question. An initial lag phase during which the herbicide concentration remains constant is followed by a period of rapid disappearance (curve A, Figure 12.11). Enzymes specific for the metabolism of the foreign molecule are induced in responsive soil organisms, principally bacteria, during the lag phase; these adapted organisms enjoy a competitive advantage for substrate over non-adapted organisms, resulting in their proliferation and the breakdown of the herbicide. The soil is then said to be *enriched* with adapted organisms, a condition persisting for several months in the absence of fresh herbicide. Herbicide reapplied to an

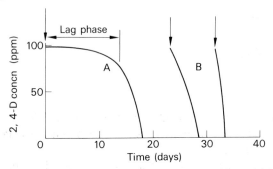

Figure 12.11. Time course of microbial decomposition of 2,4-D in soil (curve A—initial enrichment, curve B—addition of herbicide to enriched soil; arrows indicate time of addition of the herbicides) (after Audus, 1960).

enriched soil is detoxicated without any lag phase (curve B, Figure 12.11).

Apart from complete detoxication, many soil organisms can partially degrade pesticides that are analogues of normal substrates, a process called *co-metabolism*. An organism adapted to benzoates, for example, may oxidize a chlorobenzoate herbicide to a catechol derivative which is more prone to oxidation by the general soil microflora than the original compound.

12.6 SUMMARY

Substances recognized as fertilizers and pesticides have been used for thousands of years. But in the short period after World War II there has been a phenomenal increase in the quantity and variety of agricultural chemicals to meet the need for greater production per hectare and improved quality of the crops harvested.

The main nutrients supplied are N, P and K either singly in *straight* fertilizers or in various combinations in *mixed* or *compound* fertilizers. Most of the simple compounds of N—NH_4NO_3, $NaNO_3$, $Ca(NO_3)_2$, $(NH_4)_2SO_4$ and $(NH_2)_2CO$—are soluble solids, but N is also used as solutions of NH_4NO_3, urea or NH_3 in water and as liquefied NH_3 gas. The N of these compounds is immediately available to the plant, but is vulnerable to leaching in the NO_3 form and losses range from 10–100 kg N ha^{-1} an^{-1} depending on the rate of fertilizer applied, the land use and the excess winter rainfall. The main K compounds—KCl and K_2SO_4—are also very soluble, although K^+ is retained as an exchangeable cation in most soils.

The naturally occurring phosphate minerals are most insoluble and must be treated with strong acids (H_2SO_4, HNO_3 and H_3PO_4) to convert the bulk of the P to a soluble form. Monocalcium phosphate, $Ca(H_2PO_4)_2$. H_2O, is the active constituent of single and triple *superphosphates*, while higher analysis P fertilizers (the *polyphosphates*) are made from superphosphoric acid which is formed by reducing the amount of water in which P_2O_5 is dissolved. Soluble P fertilizers react quickly with the soil to form less soluble *fertilizer reaction products* on which plants feed.

Ideally, a *pesticide* should be lethal for a specific organism and remain active only long enough to control hat pest. Pesticide chemicals are inactivated in the soil by adsorption onto clays or organic matter, by microbial enzymic attack or chemical decomposition. Of the more stable compounds, the organochlorine insecticides have become widely disseminated in the environment due to volatilization from the soil and their accumulation in animal fatty tissues. Other more soluble compounds like the phenoxyacid and urea herbicides may be leached, but fortunately most are decomposed by microorganisms.

REFERENCES

LARSEN S. (1971) Residual phosphate in soils, in *Residual Value of Applied Nutrients. MAFF Technical Bulletin* **20**, 34–40.

FURTHER READING

AGRICULTURE AND WATER QUALITY (1976) *MAFF Technical Bulletin* **32**.

COOKE G.W. (1975) *Fertilizing for Maximum Yield*. 2nd Ed. Crosby Lockwood, London.

GREEN M.B., HARTLEY G.S. & WEST T.F. (1977) *Chemicals for Crop Protection and Pest Control*. Pergamon, Oxford.

GUENZI W.D. (1974) (Ed.) *Pesticides in Soil and Water*. Soil Science Society of America, Madison.

GUNN D.L. & STEVENS J.G.R. (1976) (Eds.) *Pesticides and Human Welfare*. Oxford University Press, Oxford.

HUFFMAN E.O. (1968) The reactions of fertilizer phosphate with soils. *Outlook on Agriculture* **5**, 202–207.

NITROGEN AND SOIL ORGANIC MATTER (1969) *MAFF Technical Bulletin* **15**.

SLACK A.V. (1967) *Chemistry and Technology of Fertilizers*. Wiley Interscience, New York.

Chapter 13
Problem Soils

13.1 A BROAD PERSPECTIVE

Nutrient deficiencies, soil acidity, structural instability or a soil's susceptibility to erosion can restrict the growth of natural vegetation or agricultural crops: these problems have been explored in the preceding three chapters. Excess water in the soil can also severely limit the use of land for agriculture.

As discussed in section 9.2 and 9.5, an excess of precipitation over evaporation for several successive months each year, impermeable subsurface layers and high groundwater tables separately or together induce *waterlogging* and the attendant problems of
(1) inadequate aeration for root and soil microbial activity;
(2) poor trafficability of the soil for machinery and animals, with the danger of surface structural collapse and "poaching"*;
(3) invasion by flood-tolerant weeds and an increase in animal parasites and diseases that are favoured by wet conditions;
(4) slow warming of soil in spring.

The causal climatic factors are beyond control, but an improvement in soil drainage (section 13.4) does much to mitigate these ill effects.

On the other hand, in areas of high evaporation, serious problems occur when the water table rises to within 1 to 2 m of the soil surface and *salinization* occurs due to the upward movement and evaporation of saline groundwater (section 6.6). Such a situation may arise when the steady-state equilibrium of the soil's hydrology is disturbed by faulting in the Earth's crust or volcanic activity (measured in a time span of thousands of years), or by man's intervention through changing the land use or supplying irrigation (usually a time span of tens of years). Problems associated with hydrological disturbances, and their solutions, fall into two categories

(a) soils not initially saline with deep watertables which become saline under irrigation or due to a change in land use (section 13.2).
(b) soils already salinized and of limited cropping potential but requiring careful management to avoid chemical and physical deterioration under irrigation (section 13.3).

13.2 WATER MANAGEMENT FOR SALINITY CONTROL

Permeation of saline groundwater
The existence of large salty lakes and surrounding salt pans in many inland areas is evidence of the confluence of natural drainage in depressions in which salts accumulate following the evaporation of water. In the southwest of Western Australia, many of the valley bottoms are naturally salinized and support a scrubby heath vegetation tolerant of high salinity. Groundwater levels are kept low under the valley slopes by the deep-rooted indigenous Eucalyptus trees (Figure 13.1a), but following forest clearance for arable farming, the rate of deep percolation increases from some 5 to 10 mm yr^{-1} and the watertable rises in the lower slopes due to the increased hydraulic gradient. Thus the area of saline soils creeps gradually upslope (Figure 13.1b).

Soils under irrigation

REGULATING THE WATER SUPPLY
Some soils, notably those of the Nile Valley, have been irrigated for centuries without ill effect. When the Nile flooded in August and September, the water was led onto the land and impounded in basins of 400–16,000 ha for up to 40 days, after which the water receded and crops were planted. The soils were naturally well drained and the annual flooding leached out any accumulated salts. More recently, in order that irrigated crops could

* Rutting and damage to the wet soil surface by animals

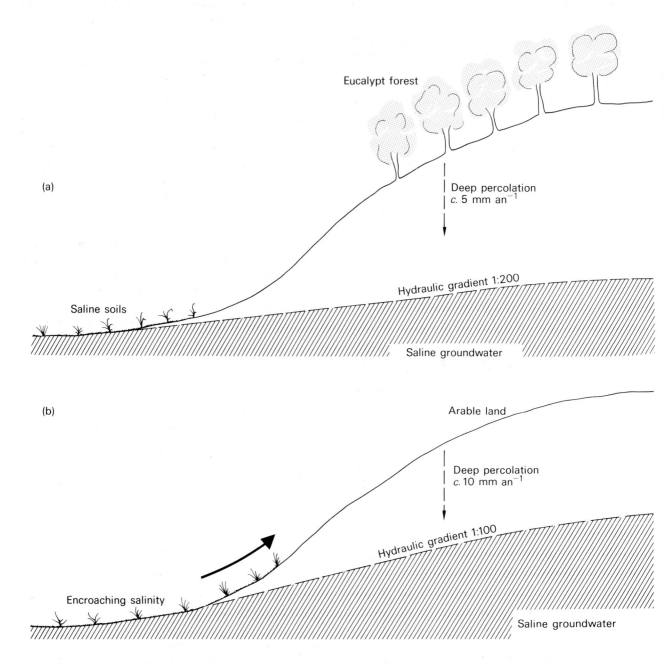

Figure 13.1. (a) Natural landscape in south-western Australia in steady-state hydrological equilibrium. **(b)** Same landscape with raised groundwater table after forest clearance (after Holmes, 1971).

be grown throughout the year, dams and barrages have been built to regulate the river's flow. Productivity has increased but not without cost, for with the application of 1.5 to 2 m of water annually, most of which is evaporated or transpired by the crop, there has been an inevitable build-up of salts in the soils.

Abundant and regular supplies of water may also encourage farmers to overwater, the excess water contributing to the groundwater which slowly rises. Another hazard is the seepage loss from canals and ditches which conduct water to the fields: this problem is especially serious in Pakistan where extensive water-logging of the permeable soils of the Indus Valley occurs alongside the irrigation canals which are long and very wide (Figure 13.2).

Figure 13.2. Soils waterlogged by lateral seepage from a large irrigation canal in Pakistan (courtesy of H. van Someren).

Seepage is prevented by lining canals with concrete— the best but most expensive material; asphalt or plastic sheeting are cheaper but of limited durability. Water use must be adjusted according to the measured or calculated *soil moisture deficit* (*SMD*). The depletion of available water can be calculated from soil moisture measurements made with gypsum resistance blocks or neutron probes. More appropriate for large areas is the prediction of *SMD* from evapotranspiration losses (*ET*) calculated from the Penman equation (section 6.6). Thus,

$$ET \text{ (mm day}^{-1}) \times \text{days from last watering} =$$
expected *SMD* per metre depth

Given that available water capacities (*AWC*) normally range from *c.* 80 mm m^{-1} for sandy soils to 200 mm m^{-1} for silty clays and clay loams, an *ET* rate of 6 mm day^{-1} (see Figure 6.14) would remove just over half the available water in 7 to 17 days, depending on the soil texture, at which point it would be desirable to irrigate again to avoid a check to crop growth.

IRRIGATION METHODS

The method of applying water also influences the likelihood of soil salinization. Near level or uniformly graded land can be irrigated by *flood* or *furrow* methods. *Sprinklers*, *sprays* or *trickle emitters* are suitable for irregular terrain.

(a) With *border check* or *border strip* flooding, water is distributed along bays separated by levees or mounds running parallel to the direction of flow (Figure 13.3)

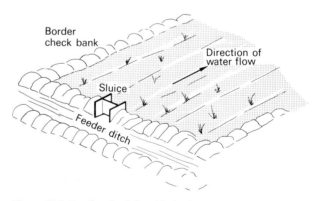

Figure 13.3. Border check flood irrigation.

The method is suitable for slopes from 4 to < 1 per cent. Sufficient depth of water (> 50 mm) must be applied at each irrigation to ensure that some water reaches the bottom end of the bay before infiltrating the soil. To avoid overwatering at the input end, the bays should be shorter in permeable soils (< 90 m) than in less permeable clays (up to 270 m). On the lowest gradients (< 1 per cent), construction of levees across the slope at vertical intervals of approximately 5 cm to form rectangular basins gives more uniform water application, and therefore better control of salinity. However, this method of *basin flooding* is more labour intensive and usually practical only on highly productive orchard and padi rice crops.

(b) *Furrow irrigation* is obviously suited to row crops and can be used on land too steep or uneven to be flooded, provided the furrows are keyed to the contours. But there is the problem of salt redistribution in the soil as water moves by capillarity from the wet furrows to evaporate from the ridges where the plants are growing. The salt concentration at the apex of the ridge may be 5 to 10 times higher than in the body of the soil, creating an unfavourable environment for seed germination and young seedling growth. Measures to avoid this problem consist of sowing seeds in double rows on broad sloping ridges (Figure 13.4a) or in single rows on the longer slope of asymmetric ridges (Figure 13.4b).

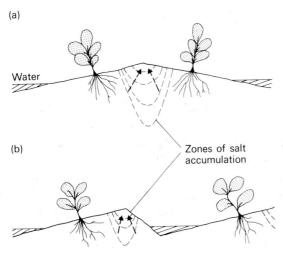

(a)

Water

(b)

Zones of salt accumulation

Figure 13.4. Ridge and furrow designs to avoid salinity damage to plants. **(a)** Paired crop rows on broadly sloping ridges. **(b)** Single crop rows on asymmetric ridges.

(c) *Spray irrigation* allows greater flexibility in the use of terrain and depth of water applied than either flood or furrow methods, but it requires more costly equipment. The spray pipes are fixed, as in orchards and market gardens, or moved manually; more recently

self-propelled water guns able to spray a circular area of *c*. 1 ha have become available. A drawback is the evaporation loss, especially on windy days, so that too little water is applied and leaching is inadequate to maintain a salt balance. Evaporation of water from the foliage can leave salt deposits and cause leaf scorch.

Trickle or *drip irrigation* is an adaptation of spray irrigation that minimizes evaporative losses and is especially suited to high-value orchard and row crops on soils of low *AWC*. Water is delivered through flow regulating 'emitters' which are placed close to the plants at regular intervals. Adjustment of the flow rate to the transpiration of the crop and the soil's permeability ensures that only enough water is supplied to keep the active root zone at a high water potential. Adverse salt effects are avoided because the salt concentration of the soil water remains comparable to that of the irrigation water; but salt accumulation at the fringe of the wet soil volume may necessitate periodic flushing of the whole soil if salinity problems for subsequent crops are to be avoided.

Quality of irrigation water

KINDS OF SALTS AND THEIR CONCENTRATION
The major ions in surface and underground waters used for irrigation are Ca^{2+}, Mg^{2+}, Na^+, Cl^-, $SO_4^=$ and HCO_3^- with small amounts of K^+ and NO_3^-. The total concentration of dissolved salts (*TDS*), sometimes referred to as the 'salinity hazard', is most conveniently measured by the specific conductance, which is the electrical conductivity (*EC*) of the water, independent of the sample size. Analysis of many surface and well waters reveals a linear relation between log *TDS* in ppm and log *EC* in μmhos cm^{-1} such that an *EC* of 1000 at 25°C is equivalent to 640 ppm *TDS* (Figure 13.5).

Experience in the western U.S.A. (Richards, 1954) suggests that for waters of *low, medium, high* and *very high* salinity, the *EC* class limits are <250, 250–750, 750–2250 and >2250 μmhos cm^{-1} at 25°C. In the absence of salt accretion from the groundwater, the concentration of the soil's saturation extract is usually 2 to 3 times greater, depending on the depth, than the concentration of the irrigation water, so that the continued use of waters of high salinity is likely to create

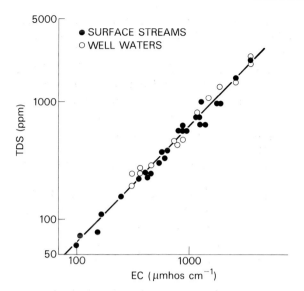

Figure 13.5. Relationship between *TDS* and electrical conductivity for surface and well waters (after Richards, 1954).

a *saline* soil, defined as one in which the *EC* of the saturation extract (EC_e) is 4000 μmhos cm^{-1} or greater.

SODIUM HAZARD

Apart from total salinity, the content of boron and sodium is likely to affect the suitability of water for irrigation. Boron concentrations > 1 ppm are potentially toxic to plants, and if Na is high relative to Ca and Mg, structural degradation of the soil may occur (section 9.6). The Gapon equation, introduced in section 7.2, is valuable for quantifying the relation between the activity ratio Na/$\sqrt{\mathrm{Ca+Mg}}$ in solution and the ratio of Na to (Ca+Mg) on the exchange surfaces. For a three cation system, the equation may be written as:

$$\frac{\text{exchangeable Na}}{CEC - \text{exchangeable Na}} = k_{\mathrm{Na-Ca,Mg}} \frac{(\mathrm{Na})}{(\mathrm{Ca+Mg})^{\frac{1}{2}}} \quad (13.1)$$

where the quantities on the LHS are in me per 100 g soil and the concentrations on the RHS in molarities.

As irrigation water is concentrated in the soil by evapotranspiration, assuming insoluble salts are not precipitated, the concentrations of all the ions will increase in the same proportion. For a twofold increase in the ionic concentrations, however, the value of

Na/$\sqrt{\mathrm{Ca+Mg}}$, which is called the *sodium adsorption ratio (SAR)*, increases by $2/\sqrt{2}=1.4$ so that exchange occurs to raise the proportion of Na relative to (Ca+Mg) on the clay surfaces. The RHS of equation 13.1 therefore increases as the *SAR* of the irrigation water increases, from which it follows that there is an almost linear rise in the value of

$$\frac{\text{exchangeable Na}}{CEC} \times 100 = \text{Exchangeable Na percentage}$$
$$(ESP) \quad (13.2)$$

with increase in the irrigatio nwater *SAR* (Figure 13.6).

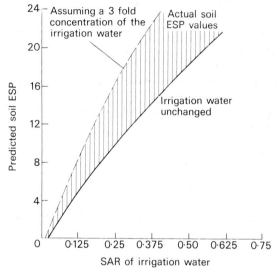

Figure 13.6. Relationship between soil *ESP* and *SAR* of irrigation water (after Richards, 1954).

Prediction of the soil's *ESP* is vital to irrigation management because swelling pressures are enhanced as the *ESP* increases, especially if the salt concentration of the soil solution is subsequently lowered. The result is a reduction in the soil's macroporosity and permeability (section 13.3).

ESP values actually found in surface soils are slightly higher than those predicted by the solid curve in Figure 13.6 because of the concentrating of the irrigation water once it enters the soil. The dotted line in Figure 13.6 gives the *ESP–SAR* relationship assuming a three fold rise in the total concentration of the irrigation water in the soil. Moreover, even for waters of the *same SAR*, the sodium hazard is deemed to be greater, the greater the total salinity. This may be caused by precipitation of Ca and Mg salts (CaSO$_4$, MgCO$_3$) from a high TDS water as it becomes more concentrated; the removal of Ca and Mg from solution before Na results in the SAR increasing faster than is predicted for a given increase in the total solution concentration, and the ESP is correspondingly higher.

LEACHING REQUIREMENT

Even good quality irrigation water contains some dissolved salts and because the crop demands very much more of the water than the salts, any unused water becomes more salty as it percolates through the soil. To prevent salts accumulating, sufficient water must be applied to maintain drainage out of the soil at a concentration below the maximum tolerated by the particular crops grown. In the absence of salt precipitation and ignoring rainfall, a favourable salt balance will be maintained if

$$EC_{IW}\, d_{IW} = EC_{DW}\, d_{DW} \qquad (13.3)$$

where EC_{IW} and EC_{DW} are the electrical conductivities of the irrigation and drainage waters, repectively; and d_{IW} and d_{DW} are the corresponding water depths summed over the year. It follows that

$$\frac{EC_{IW}}{EC_{DW}} = \frac{d_{DW}}{d_{IW}} = \text{Leaching requirement} \qquad (13.4)$$

The leaching requirement (LR) can therefore be determined once the crop's salinity tolerance in terms of EC_e values is known (Table 13.1).

Table 13.1. Crop tolerance to salinity as measured by the EC_e of the soil.

EC_e (mmhos cm^{-1}) at 25°C	Crop response
0–2	Negligible effects
2–4	Yields of very sensitive crops reduced e.g. beans, clovers
4–8	Yields of many crops reduced
8–16	Only tolerant crops not seriously affected e.g. Bermuda grass, sugarbeet
>16	Only a few very tolerant crops grow satisfactorily e.g. barley

(after Bernstein, 1970)

The total depth of irrigation water required to satisfy the crop's consumptive use (d_{CW}) and the LR is calculated by rewriting equation 13.3 in the form:

$$EC_{IW}\, d_{IW} = EC_{DW}\,(d_{IW} - d_{CW})$$

from which

$$d_{IW} = \frac{d_{CW}}{1 - LR} \qquad (13.5)$$

13.3 RECLAMATION OF SALT-AFFECTED SOILS

Permeability changes on leaching

A saline soil becomes a *saline–sodic* soil when the ESP exceeds 15 per cent. Leaching of a saline–sodic soil may reduce the salt content such that the EC_e value of the soil falls below 4 mmhos cm^{-1}, when it is classed as *sodic*. Both classes of soil need very careful management because of their high ESP values. As Figure 13.7 shows,

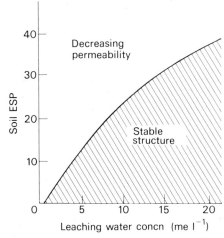

Figure 13.7. Interactive effect of ESP and concentration of the leaching water on soil permeability (after Quirk and Schofield, 1954).

for a given ESP value there is a threshold concentration of leaching water below which the soil permeability decreases markedly due to the swelling of clay domains and subsequent clay deflocculation. Thus, the lower the electrolyte concentration of the soil solution, the lower the ESP value at which the structure becomes unstable

Reclamation of saline, saline–sodic and sodic soils therefore requires leaching with water of a low enough SAR to initiate Ca^{2+} exchange for Na^+, but a sufficiently high total salt concentration to preserve the soil's permeability. As the ESP is gradually reduced, the salinity of the leaching water may also be reduced until reclamation is complete.

SOURCES OF CA²⁺ IONS

Calcium is usually supplied as gypsum which has a solubility of 30 me 1^{-1} and can be spread on the soil or dissolved in irrigation water. Approximately 3.4 t is required to reduce by one unit the ESP of a volume of soil one ha by 15 cm deep. Calcium carbonate, if present, is normally too insoluble to lower the SAR of the leaching water significantly; but in soils containing *Thiobacillus* bacteria, the addition of sulphur enhances the solution of $CaCO_3$ due to the acidity generated when S is oxidized to $SO_4^=$, according to the reaction:

$$S + \frac{3}{2}O_2 + H_2O \rightarrow 2H^+ + SO_4^= \qquad (13.6)$$

Metal sulphides in marine sediments behave similarly when first exposed to air (section 11.3), a reaction to the benefit of reclamation in Dutch polders, where the H^+ and SO_4 ions react with indigenous $CaCO_3$ to produce gypsum sufficient to saturate the soil solution for several years as the reclaimed land is leached by rainwater.

BLENDING OF LOW AND HIGH SALINITY WATERS

The theory of saline soil reclamation was put to an interesting test with a saline–sodic soil from the Coachella Valley in the U.S.A. The soil, which had an ESP of 39 and an EC_e of 4.4 mmho scm^{-1}, was leached with Salton seawater that was progressively diluted with good quality Colorado riverwater. Each water mixture, of which the range in composition is given in Table 13.2, was leached through the soil until the soil was in equilibrium with that mixture. Using four stepwise reductions in the leaching water concentration, the soil permeability did not fall below 0.12 m day^{-1} and the ESP was lowered to 5 in only 12 days. But if the soil was leached with Colorado riverwater alone, the permeability fell to 0.005 m day^{-1} and reclamation took 120 days.

Table 13.2. Reclamation of a saline–sodic soil with water of stepped-down salt concentration and SAR.

Dilution ratio (seawater + riverwater)	Water properties		Soil properties	
	TDS (me l^{-1})	SAR (me$^{\frac{1}{2}}$ l$^{-\frac{1}{2}}$)	Hydraulic conductivity (m day^{-1})	ESP %
Salton Sea	562	57	—	—
1 : 3	149	27	0.13	28
1 : 15	45.4	12	0.12	14
1 : 63	19.6	5	0.11	6
0 : 1	11.0	2	0.12	5
Colorado River	11.0	2	0.005	5

(after Reeve and Bower, 1960)

Removal of surplus water

It is corollary of good irrigation management that surplus water, arising out of a leaching requirement or through seepage to a shallow groundwater table, should be disposed of efficiently. In permeable soils, the natural hydraulic properties may be able to cope with the excess water; where this is not the case, soil drainage must be improved by the installation of underground porous pipes or *tile drains* through which water is rapidly led away.

13.4 SOIL DRAINAGE

Choosing a drainage design

Irrigated soils in dry climates, for which the depression of the groundwater table and the disposal of water in excess of the crop's needs are prime objectives, require a drainage design philosophy different from soils of cool humid climates where the natural permeability may be inadequate to prevent the soil becoming waterlogged for unduly long periods (1 or more days) when precipitation exceeds evaporation. Irrigation schemes are also very capital-intensive, producing high returns per hectare, so that more expensive drainage systems may be justifiable on irrigated than on non-irrigated land. These two broad categories of problem soils therefore warrant discussion individually.

IRRIGATED SOILS

The variables which interact to determine the equilibrium height of the watertable in a soil with tile drainage are:

(a) The hydraulic conductivity k of the saturated soil (section 6.3);

(b) The depth of drains d and their height h above any impermeable barrier in the soil, that is, a layer with a k value ≤ 0.1 of the layer above;

(c) The distance between drains L;

(d) The mean rate of water percolation q through the soil to the groundwater; to avoid salt accumulation in the soil, this rate should be at least 1 mm day^{-1}.

Flow to tile drains is horizontal as well as vertical. For horizontal flow, the watertable must rise between the drains to maintain a hydraulic gradient, as illustrated in Figure 13.8. To preclude capillary rise of saline groundwater, the watertable should be at least 2 m

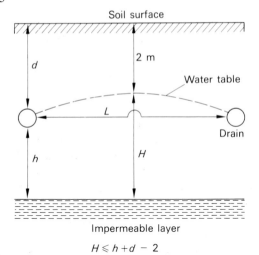

$$H \leqslant h + d - 2$$

Figure 13.9. Inter-relations between variables which determine minimum drain spacings.

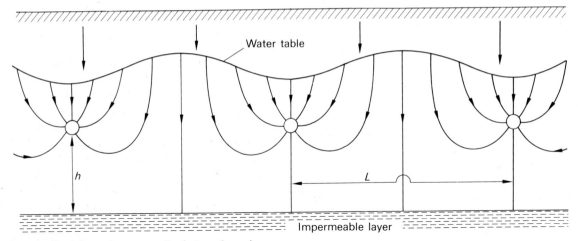

Figure 13.8. Idealized flow of water to tile drains after rain or irrigation.

below the soil surface when the drains have stopped flowing. But allowance must be made for the elevating effect of an impermeable layer on the equilibrium watertable position *between* the drains when it occurs at a depth $h < 0.2 L$. The drain spacing necessary to keep

$$H \leq h + d - 2 \text{ (see Figure 13.9)}$$

where all depths are in metres can be calculated from the equation

$$L = \sqrt{8 \, i \, (H-h) \, k/q} \qquad (13.7)$$

using measured values of k, in m day^{-1}, and $q = 0.001$ m

day^{-1}. The term i is an index value dependent on the values of L, h and the tile drain diameter, so that the equation must be solved iteratively since factors affecting i occur as variables in the question. However, the solution is much simplified by the use of nomograms devised from the analysis of field data (Visser, 1945).

In the case of many clay soils, where the impermeable layer lies at or just below drain depth ($h \simeq 0$), deep flow through the soil to the drains is inhibited and equation 13.7 reduces to

$$L = 2H\sqrt{k/q} \qquad (13.8)$$

Often, however, the calculation of a theoretical drain spacing is vitiated by the great spatial variation in k so that predictions based on a uniform flow pattern, as in Figure 13.8, must be tempered by experience and practical judgement.

SOILS OF HUMID REGIONS

A widely accepted design criterion for draining wet impermeable soils is to lower the watertable to 0.5 m below the surface 24 hr after the end of rainfall; this will create a moisture suction of 50 mbars at the surface, corresponding to the *FC* of a well drained soil. There are two major constraints on the effectiveness of drainage in such soils:

(1) The *pore volume* that is drained at 50 mbars suction, for this determines the maximum air-filled porosity of the drained soil; this value depends on the moisture characteristic curve of the soil, but if less than 10 per cent, the improvement in aeration which drainage can achieve will be very limited.

(2) The *hydraulic conductivity* of the saturated soil which governs the rate of drainage at whatever head gradient can be applied. The conductivity of clay soils in Britain ranges from 10–100 m day^{-1} in the dry state, when deep fissures are prominent, to 0.01–0.001 m day^{-1} when saturated. Accepting that the drainage system should be capable of removing 10 mm of surplus water per day, one can calculate that, for an impermeable layer varying from zero to an infinite depth below the drains, the theoretical drain spacings for a soil of $k = 0.01$ m day^{-1} are as little as 1 to 2.2 m. Such close tile drain spacings are rarely economically feasible.

Underdrainage in practice

DRAIN DEPTH

In order that the watertable in quiescent periods is nowhere within 50 cm of the surface, tile drains need to be placed more deeply, the actual depth depending on the position of any impermeable layer and the adequacy of the outfall into the collection ditch. The drains are laid with a gentle gradient at a depth between 70 and 120 cm. The deeper the drains the drier the soil in the plough layer at equilibrium, but this condition is not

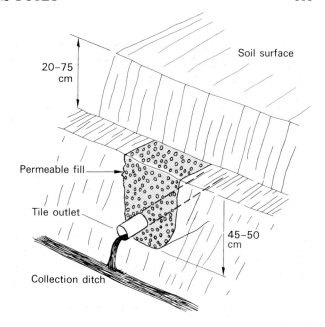

Figure 13.10. Diagram of a tile drain imbedded in permeable fill.

always attained in clay soils in winter owing to their low saturated permeabilities.

Tile drains are either short lengths of porous clay pipes, usually 75 mm in diameter, laid end-to-end, or continuous plastic pipes 50 mm in diameter which are corrugated for strength and perforated to allow water entry. Because water enters the pipe from all sides, the drain's performance is much improved if it is imbedded in a *permeable filling material* (Figure 13.10). Permeable fill, usually consisting of gravel or crushed stone of size range 5 to 50 mm, is of benefit for three reasons:

(a) It filters out fine soil particles which may otherwise block the drain.

(b) It reduces the loss in head caused by the restricted number of entry points for water into the drain.

(c) It serves as a connexion between the tile drains and any mole drains which are drawn at shallower depths.

DRAIN SPACING

Since laying tile drains at the spacing necessary to drain effectively soils of hydraulic conductivities as low as 0.1 to 0.001 m day^{-1} is clearly very expensive, the practical alternative is to choose an economic spacing—between 20 and 80 m but commonly 20–40 m—and to try and increase water flow to the drains by secondary treatments, of which *moling* and *subsoiling* are the most important.

Moling provides cheap drains at the spacing required for efficient drainage in impermeable soils. The depth of the mole drains determines the equilibrium watertable position and the spacing governs its rate of rise and fall during and after rainfall. Closer spacings delay the rise and hasten the fall of the watertable.

Subsoiling is intended to improve the hydraulic conductivity of the soil to the point where a drain spacing of 20–40 m provides efficient drainage. As equation 13.8 shows, however, a 10-fold increase in drain spacing must be complemented by a 100-fold increase in k, which is not often achieved because the soil at depth is too moist for effective subsoiling. Nevertheless, subsoiling can be effective where a plough pan markedly reduces the vertical permeability of the soil above the drains.

Figure 13.11. Diagram of a mole drainer in operation (after MAFF Field Drainage Leaflet No. 11).

METHODS OF MOLING AND SUBSOILING

Moles are drawn by the implement illustrated in Figure 13.11. The 'bullet' of 75 mm diameter is pulled through the soil at 50–60 cm depth and a spacing of 2–3 m. Behind the bullet, the expander of 100 mm diameter consolidates the walls of the channel and seals the slit left by the vertical blade. Passage of the bullet and blade through the soil should fracture the structure and improve flow to the mole drain. Mole drains should lie at right angles to the tiles, provided their gradient can be kept between 2 and 5 per cent, and pass through the top of the permeable fill (Figure 13.11). Whereas tile drains should last 50 years and more, moles should be redrawn within 5–10 years for best results.

In subsoiling, a wedge-shape share is pulled at right angles to the tile drains as deep as 50 cm if possible. Ideally, at this depth and at a spacing of about 1.3 m, the structure of all the upper soil profile should be

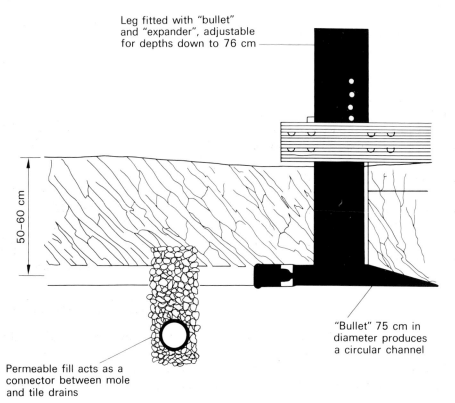

Leg fitted with "bullet" and "expander", adjustable for depths down to 76 cm

50–60 cm

"Bullet" 75 cm in diameter produces a circular channel

Permeable fill acts as a connector between mole and tile drains

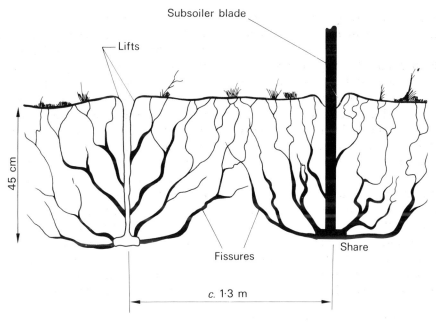

Figure 13.12. Soil fissuring induced by effective subsoiling (after MAFF Field Drainage Leaflet No. 10).

disturbed by fracturing and fissuring (Figure 13.12). The effect is more transient than moling.

The success of moling and subsoiling depends largely on the soil's *consistency* at the time of the operation. Consistency describes the change in the physical condition of the soil with moisture content, and reflects both the cohesive forces between particles (ped shear strength) and the resistance of the soil mass to deformation (bulk shear strength). With reference to the consistency diagram of Figure 13.13, it is found that moling is best when the soil is plastic, but nearer to the lower plastic limit, whereas subsoiling is more effective when the soil is friable, nearer to the shrinkage limit. In England and Wales, these conditions are generally satisfied if the SMD in the whole soil to 1 m depth is at least 50 mm (for moling) and 100 mm (for subsoiling).

13.5 SUMMARY

Excess soil water due to a consistently high P/E ratio, an impermeable subsoil or a high groundwater table imposes severe limitations on land utilization. In cool humid climates, the consequential problems of in-

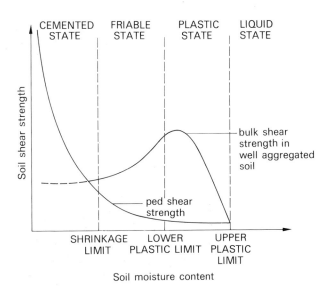

Figure 13.13. Variation in soil shear strength and consistency with moisture content (after MAFF, FDEU Technical Bulletin 75/5).

adequate aeration, poor trafficability, invasion by flood-tolerant weeds, pests and disease, and low soil temperatures in spring can be mitigated by the installation of *pipe* or *tile drains* to create a suction of at least

50 mbars at the soil surface within 24 hr of rain, provided that (1) the pore volume drained is 10 per cent or more of the soil volume and (2) the hydraulic conductivity is such that water is transmitted to the drains at a rate of at least 0.01 m day^{-1}.

If the hydraulic conductivity is so low that the tile drains would need to be placed at uneconomically close spacings (1–2 m), the practical solution is to use wider spacings (20–40 m) with secondary treatments such as *moling* and *subsoiling* to improve the flow to the tiles. Mole drains are drawn above the tiles at intervals of 2–3 m, preferably when the *SMD* is at least 50 mm in the top metre. Imbedding the tile in 40–50 cm of *permeable fill* promotes peripheral flow into the tile and provides a good connexion with the moles above. To be effective, subsoiling must be carried out when the *SMD* is at least 100 mm, but it does not last as long as moling.

Underdrainage is also essential on many irrigated soils in dry climates to prevent saline groundwater rising too near the surface. When this occurs, capillary rise and high evaporation rates lead to soil salinization, with high *ESP* values on the clay and a *potentially* unstable structure. Even with irrigation water of low *TDS* and *SAR*, salts and Na$^+$ ions accumulate in the soil with time as evapotranspiration continually concentrates the soil water. As shown by the ratio law, concentration of the soil solution favours the replacement of Ca^{2+} and Mg^{2+} by Na$^+$ on the clay. This trend can be counteracted by the flow of *c*. 1 mm of water per day below the root zone—a *leaching requirement*, which is easily

dissipated in permeable soils by deep natural drainage but which must be removed by tile drains in less permeable soils.

Reclamation of *saline* soils (*EC$_e$* of saturation extract > 4 mmhos cm^{-1}) and *saline-sodic* soils (*EC$_e$* > 4 mmhos cm^{-1} and *ESP* > 15) must be undertaken with water of salt concentration high enough to maintain soil permeability, but of low enough *SAR* to promote the displacement of exchangeable Na$^+$ by Ca^{2+} and Mg^{2+}.

REFERENCES

RICHARDS L.A. (1954) (Ed.) Diagnosis and improvement of saline and alkali soils. *United States Department of Agriculture, Handbook No.* **60**.

VISSER W.C. (1954) Tile drainage in the Netherlands. *Netherlands Journal of Agricultural Science* **2**, 69–87.

FURTHER READING

ALLISON L.E. (1964) Salinity in relation to irrigation. *Advances in Agronomy* **16**, 139–180.

CHILDS E.C. (1970) Land drainage: an exercise in physics. *Outlook on Agriculture* **6**, 158–165.

LUTHIN J.N. (1957) (Ed.) *Drainage of agricultural lands.* American Society of Agronomy, Madison.

TALSMA T. & PHILIP J.R. (1971) (Eds.) *Salinity and Water Use.* Macmillan, London.

THOMASSEN A.J. (1975) (Ed.) Soils and field drainage. *Soil Survey of England and Wales, Technical Monograph No.* **7**.

Chapter 14
Soil Survey and Classification

14.1 THE NEED TO CLASSIFY SOILS

Soil variability

The point was made in Chapter 1 that the distinctive character of a soil is shaped by a complex interaction of many physical, chemical and biotic forces. This theme was expanded in subsequent chapters on organic matter, structure formation, water movement, reactions at surfaces and so on; but a study of soil in the field makes it clear that the balance of forces, and the speed with which they act, vary from site to site according to the local expression of the soil-forming factors, as discussed in Chapter 5. Thus, the many *morphological* properties (e.g. colour, texture, structure, horizon development) and *edaphic** properties (e.g. pH, *CEC*, organic C) which characterize a soil display a very wide range in value. Man, in exploiting the soil, also alters the balance of natural processes and so adds to the total pool of *soil variability*.

The existence of soil variability must influence decisions made by potential users at various levels of endeavour:

(a) At the national level, in attempting to reserve for urban, industrial, agricultural and recreational development those areas of land best suited to their respective needs;

(b) At the town and regional level, in the allocation of land for residences, parks, waste disposal, roads and 'green belts';

(c) At the farm level, in the choice of which fields to plough, to put down to improved pasture or to keep for rough grazing;

(d) In the research laboratory, where the scientist works with small samples of soil, from a few grams to kilograms, assumed to be representative of a much larger volume in the field.

* Properties relevant to plant growth.

The purpose of classification

Faced with variability within natural populations, our natural response is to try and impose an ordered structure on the fund of knowledge derived from our experience with individuals in the population. This usually involves three steps:

(1) Identification of the full range of variability within the population;

(2) Creation of groups or *classes* within the population according to the similarity between individuals;

(3) Prediction of the likely behaviour of an individual from a knowledge of the characteristics of the class as a whole.

Step (2)—the creation of classes—is called *classification*. Soil classification is more difficult and contentious than that of other natural populations because a soil lacks the hereditary characteristics by which individuals within one generation are distinguished, and which are transmitted from one generation to the next. The soil population presents a *continuum* of variation so that arbitrary judgements are inevitable in the creation of classes.

DEFINITION OF A SOIL ENTITY

Soil variability can be crudely partitioned according to the distance over which discernible changes in properties occur. Differences associated with soil faunal and microbial activity and the distribution of pores, for example, achieve maximum expression within 1 m (*short-range variation*). Since as much as half of the total variation in any one property can occur as short-range variation, it is impossible to define rigorously a fundamental *soil unit* or *entity*. Attempts to do so have resulted in the introduction of esoteric terms such as the 'pedon' and 'pedounit'; but whatever the theoretical merits of these terms, the man in the field *describes* a soil on what he sees of the profile in a pit (usually 1×0.5 m in area, dug to the parent material or to 1.5 m, whichever is the shallower), or exposed in a cutting or quarry face.

Further, he usually *classifies* the soil on the evidence of samples obtained by augering from the surface downwards (Figure 14.1).

Figure 14.1. A selection of augers and core samplers used in soil survey.

KINDS OF CLASSIFICATION
Another problem confronting the soil classifier is the choice of soil properties on which to separate his classes. Many properties can only be evaluated qualitatively—soil colour, structure, type of humus and degree of gleying for example. Other properties, such as the pore size distribution and the content of organic matter, clay or calcium carbonate, are difficult to measure. Practical constraints therefore prevent all the soil properties being used—some are selected as the *definitive* properties for class separation, the choice of which depends on the aim of the classification, discussed below.

(a) *Specific or single-purpose classification.* This kind of classification is made with a specific aim in mind and is based on one or a very few soil properties. The texture classification of Figure 2.3 and the classification of soils for droughtiness and susceptibility to waterlogging (Figure 11.5) are two cases in point. There are many other examples, especially relating to plant growth, such as the *ESP* and salinity (EC_e) status of irrigated soils, and the assessment of 'available' N, P and K levels by soil chemical tests (section 11.2). When classification is based on one or two properties, classes may be separated by subdividing the range of variation in the individual property at the minima in the frequency distribution (Figure 14.2a), or according to the user's requirements (Figure 14.2b).

(b) *General-purpose classification.* Since this kind of classification is intended for many uses, foreseen and unforeseen, the classes should be defined on as many properties as possible. Provided that the properties chosen are *relevant* to the utilization of the soil, the classification will be useful; if not, the classification may be academically elegant but of little practical value. But where the important soil properties are those requiring tedious and expensive laboratory analysis, it is expedient to choose for the separation of classes certain *diagnostic* properties—not necessarily relevant to soil use but easily assessed in the field—on the assumption (or knowledge) that they are correlated with the important properties. For example, the colour of the A horizon may be correlated with organic matter content, and mottling of the subsoil indicative of a low hydraulic conductivity. Where such natural grouping or 'clustering' of soil properties exists, the classes created should be more meaningful to the general-purpose user than those defined by arbitrary criteria.

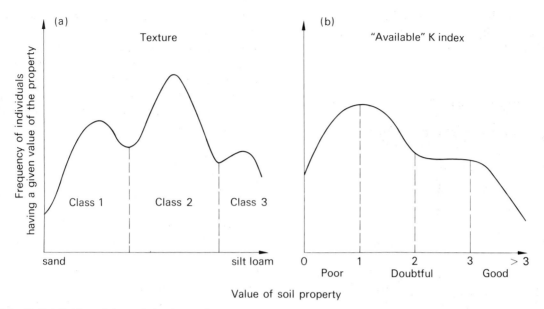

Figure 14.2 (a, b). Subdivision of the variation in a soil property to define classes for a special-purpose survey.

14.2 A GUIDE TO SOIL SURVEY

Defining the aims of survey

Sampling to assess the full range of variability and describing the soil at known locations within a target area are the first stages of *soil survey*. This kind of information is required so that informed decisions on land use, and the management best suited to different facets of land use, can be made. The effort put into survey and the procedures adopted are determined by the size of the area in relation to the human and financial resources available, and the intensity of the proposed land use. The output of survey is usually a classification and a *soil map* which shows the distribution of the different soil classes at a scale commensurate with the density of observations on the ground. The relation between the type and purpose of the survey and the map scale is summarized in Figure 14.3. The main types of survey are as follows:

(1) *Reconnaisance* and *semi-detailed* surveys, ranging in scale from 1:1,000,000 to 1:25,000, have the same broad objectives: to provide a resource inventory for large areas (provinces, states, countries). The preliminary work relies heavily on air photograph interpretation

(API) to delineate differences in vegetation, land form and management which may be the outward expression of soil differences. Once boundaries have been drawn on the air photos, surveyors go into the field to check their significance, relocating them if necessary, and to describe the main features of the soils and environment within each landscape unit. Areas suitable for various forms of land use can be located and guidance obtained as to the most suitable type of survey to be performed at the next stage.

(2) *Detailed* and *intensive* surveys range in scale from 1:25,000 to 1:1,000 and are carried out for one of three reasons:

(a) To assess the technical and economic *feasibility* of a proposed project in a given area;

(b) To assess the preliminary work necessary for the *development* of a specified project;

(c) Within a developed area, to provide data for the solution of *management* problems.

The last mentioned surveys are usually special-purpose surveys—for example, the mapping of salt-affected soils in an irrigation area or soils subject to erosion in an area of continuous cereal cropping.

Survey scale

| 1:1,000 | 1:10,000 | 1:100,000 | 1:1,000,000 |

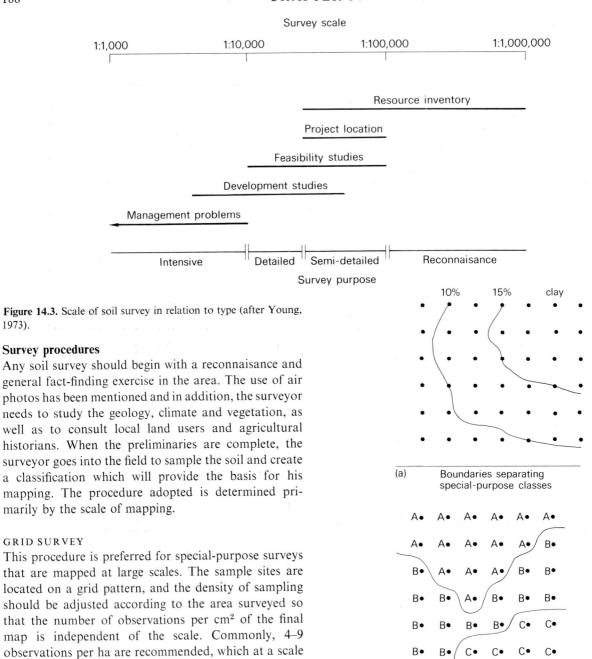

Resource inventory

Project location

Feasibility studies

Development studies

◄── Management problems

Intensive   Detailed   Semi-detailed   Reconnaisance

Survey purpose

Figure 14.3. Scale of soil survey in relation to type (after Young, 1973).

Survey procedures

Any soil survey should begin with a reconnaisance and general fact-finding exercise in the area. The use of air photos has been mentioned and in addition, the surveyor needs to study the geology, climate and vegetation, as well as to consult local land users and agricultural historians. When the preliminaries are complete, the surveyor goes into the field to sample the soil and create a classification which will provide the basis for his mapping. The procedure adopted is determined primarily by the scale of mapping.

GRID SURVEY

This procedure is preferred for special-purpose surveys that are mapped at large scales. The sample sites are located on a grid pattern, and the density of sampling should be adjusted according to the area surveyed so that the number of observations per cm² of the final map is independent of the scale. Commonly, 4–9 observations per ha are recommended, which at a scale of 1:10,000 produces 4–9 observations per cm² of the map. Boundaries are drawn to join points of equal value of a specific property, or where general-purpose classes are defined, between points at which dissimilar soils occur (Figure 14.4). Grid survey is expensive, the cost

(a)   Boundaries separating special-purpose classes

(b)   Boundaries separating general-purpose classes

Figure 14.4. (a, b) Soil boundaries drawn from grid surveys (after Beckett, 1976).

Figure 14.5. (a, b) Sampling pattern and soil boundaries drawn from free survey (after Beckett, 1976).

(a)

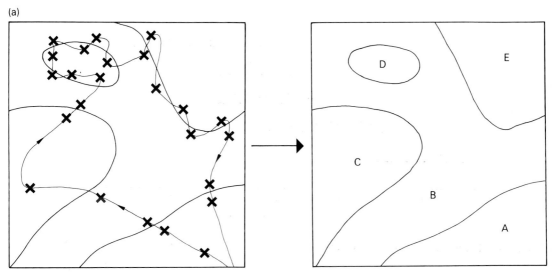

Soil surveyor's traverse (———▶) and observation sites (✖) giving the soil map on the right at a scale of 1:50,000. Boundaries of map units A and C were inferred from external features; those of units D and E were located by sampling

(b)

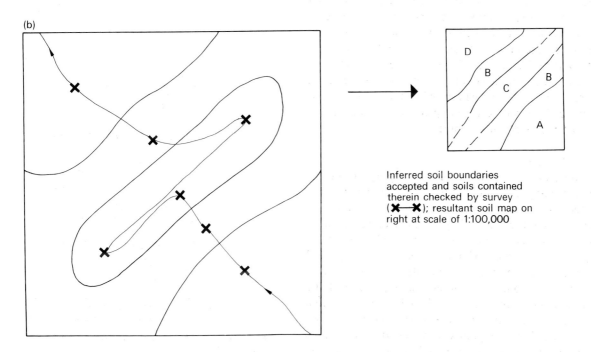

Inferred soil boundaries accepted and soils contained therein checked by survey (✖——✖); resultant soil map on right at scale of 1:100,000

increasing roughly in proportion to the square of the map scale.

FREE SURVEY
In free survey, the surveyor chooses sampling points on the assumption that changes in surface features (soil colour, relief, vegetation or land use) are indicative of soil differences. The density of sampling can be varied as the surveyor concentrates on confirming the inferred

boundaries and checking the uniformity of the soil within each boundary (Figure 14.5a). At very small scales, often the inferred boundaries must be accepted and the limited field effort directed to discovering the soils lying within the boundaries and the elucidation of any recurrent patterns, such as catenas (Figure 14.5b).

CLASSIFICATION
The surveyor's prime aim is to be able to place any soil

Figure 14.6. 'Nested' arrangement of soil mapping units.

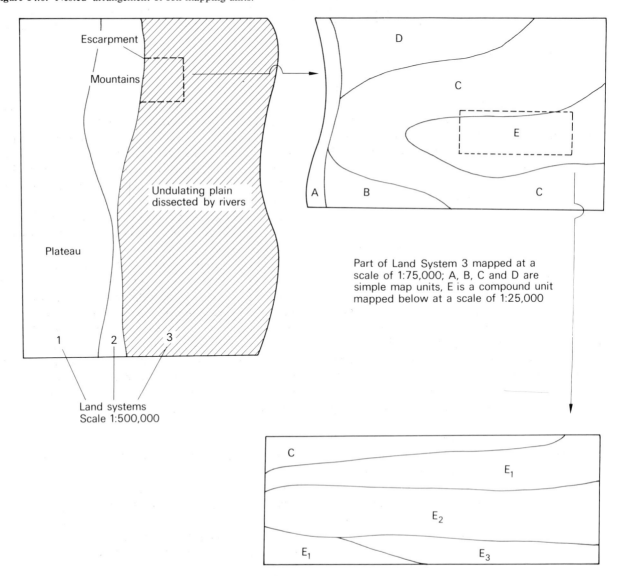

Escarpment

Mountains

Undulating plain dissected by rivers

Plateau

1   2   3

Land systems
Scale 1:500,000

D

C

A   B   C

E

Part of Land System 3 mapped at a scale of 1:75,000; A, B, C and D are simple map units, E is a compound unit mapped below at a scale of 1:25,000

C

E_1

E_2

E_1 E_3

in the area in one class. With specific-purpose classifications, the class limits are rigidly defined according to the values of one or two properties: for example, one classification of soils for irrigation purposes sets the following class limits:

saline soils $EC_e > 4$ mmhos cm^{-1}, $ESP < 15$
saline–sodic soils $EC_e > 4$ mmhos cm^{-1}, $ESP > 15$
sodic soils $EC_e < 4$ mmhos cm^{-1}, $ESP > 15$

For general purposes, however, each class is usually built around a central concept, covering a limited range in the values of several soil properties, and a certain amount of overlap in the values of any one property is permissible at the class limits.

The list of classes is the classification. Data for individual profiles is subsumed into a definitive description of each class, and one or more properties may be chosen as a diagnostic or key properties that enable an unknown soil to be quickly allocated to a class. Each class is given a locality name and referred to as a *soil series*. Soils of the same series which differ only in the texture of the A horizon are separated as *types*.

Survey results and land evaluation

SOIL MAPPING

Because all the soil profiles of any one series need not be contiguous, the distribution of soil series in the landscape is frequently intricate and necessitates some simplification for mapping. This is achieved by means of a *mapping unit*, consisting of one or more classes, which is the smallest area of the map that can be delineated by a single boundary at the scale used. At large map scales, the objective is to use *simple* mapping units, each representing a narrowly defined class (series), and to attain a purity of 80 per cent or more within that unit. But as the map scale decreases (the threshold lying between 1:30,000 and 1:75,000), or even at larger scales where a complex mosaic of classes occurs in a small area, *compound* mapping units must be used. The legend to the map lists the classes in each unit, their proportions and if possible their pattern of occurrence. At the smallest scales (< 1:500,000) complex units appear as *soil associations*, in which the classes are recognized and located first and then grouped for ease of presentation, or as *land systems* which are first distinguished by their surface expression (e.g. by API) and the included classes identified by subsequent sampling. The 'nested' arrangement of mapping units according to map scale is illustrated in Figure 14.6.

In addition to the legend, most soil maps provide a *memoir* which contains a detailed description of each soil series, with general comments about the geology and vegetation of the area and the management of soils grouped by mapping units. If the original data from the survey is recorded in a computer-compatible format, a data store can be created which is a valuable supplement to the memoir. In response to specific requests by users, selected data can be retrieved from the store and computer-printed maps produced.

LAND EVALUATION

Soil survey and classification is only one step in the complete assessment of land capability, which must also take note of the environment (climate, vegetation, topography), as well as social and economic factors which are beyond the scope of this chapter. Some argue that soil survey should go further than just classification and mapping to provide estimates of crop yields for identified soils at specified management levels (e.g. traditional, improved and advanced) and estimates of the productivity of alternative systems of land use (as net return per ha or *per capita*). This would require the collaboration of soil surveyors, agronomists and economists in analysing the results of field trials on experimental stations and records from farms on known soil types. Except for very intensive agriculture, provision of estimates on a series by series basis is unrealistic; for most purposes soil series can be grouped into *soil management units* which are defined by criteria that can be pedological (e.g. parent material) or agricultural (e.g. average slope, soil depth).

14.3 SOIL CLASSIFICATION REVIEWED

Diverse approaches to classification

When carried out as an adjunct to soil survey, classification provides the framework for making local generalizations about soil, entailing the recognition of soil series. The upwards expansion of a classification to allow progressively vaguer generalizations about the soil at higher levels or *categories* has preoccupied many individual soil scientists and corporate bodies concerned with soil survey, resulting in national or even global classifications of a hierarchical structure. Although such classifications are developed from the bottom upwards by the aggregation of small homogeneous classes to form broader groups, identification of an unknown soil proceeds stepwise from the highest category downwards.

Opinions differ widely as to the criteria to be used in establishing the different categories and for separating classes within categories. Traditionally, the soil taxonomist may choose definitive properties on the basis of
(a) the inferred *genesis* of the soil on the assumption that the soil is in equilibrium with its environment and reflects the influence of prevailing soil-forming factors; or
(b) the soil's *morphology*, being the outward expression of the processes which shaped the soil.

The success of either approach depends on the taxonomist's skill and experience in choosing definitive properties that are correlated with a large number of other soil properties. The non-traditionalist may prefer

a *numerical method* in which soil profiles are objectively sorted into classes by similarity-grouping techniques, using as many properties as possible, each given equal weight. Such classes contain maximum information but frequently do not correspond to any natural soil groupings and so are difficult to map.

The outcome of these different approaches is a profusion of classifications, some of which are summarized in Table 14.1.

But no matter how divergent current classifications are in principle, language and format, two main levels of generalization stand out: a lower level encompassing the surveyor's series which are narrow classes in harmony with the landscape, and a higher level of broad classes with somewhat overlapping boundaries, each of which embodies a central, usually pedogenetic, concept. The latter *conceptual classes* are valuable in teaching soil science because they each convey a measure of knowledge about the soil, distilled from the experience of scientists in many different lands. For this reason, the USDA classification of Baldwin, Kellogg and Thorp (1938), now unfairly derided by soil taxonomists, was presented in Table 5.3; similarly, the new World Soil

Table 14.1. Structures of a selection of national and supra-national soil classifications.

Classification	Categories
USDA Comprehensive System, 7th Approximation (Soil Survey Staff, 1960–1975)	Orders (10)* → Suborders (47) → Great Groups (185) → Subgroups (970) → Families (4500) → Series (*c.* 10,500)
World Soil Classification (FAO-UNESCO, 1974)	'World Classes' (26) → Soil Units (106)
CCTA Soil Map of Africa Classification (D'Hoore, 1964)	Main Groups (16) → Subgroups (25) → Soil Types (63)
Soil Classification for England and Wales (Avery, 1973)	Major Groups (10) → Groups (43) → Subgroups (109) → Series
Russian Soil Classification (Gerasimov and Glazovskaya, 1960)	Genetic Types → Subtypes → Species → Subspecies
Factual Key for Australian Soils (Northcote, 1960–1971)	Divisions (3) → Subdivisions (11) → Sections (54) → Classes (271) → Principal Profile Forms (855)
Canadian System (Canada Department of Agriculture, 1974)	Orders (8) → Great Groups (23) → Subgroups (165) → Families (*c.* 900) → Series (*c.* 3000)
Classification of the Australian Handbook of Soils (Stace *et al.*, 1968)	Great Soil Groups (43) → (Families) → Series
Binomial System for South Africa (MacVicar *et al.*, 1977)	Forms (41) → Series (504)

* Number of actual or potential classes within each category (after Butler, 1979)

Classification is recommended as a mechanism for the exchange of soil knowledge at an international level.

Towards a common perception of classification

The World Soil Classification arose as the list of classes in the legend to the FAO-Unesco Soil Map of the World (1974). The classification is a mono-categorical one comprising 106 classes or 'soil units'. Following the precedent of other systems, especially the USDA Comprehensive System, 7th Approximation (1960), it makes class separations on the basis of *diagnostic horizons* which are connotative of the underlying pedogenesis (just as the symbol Ea connotes a bleached subsurface horizon from which iron oxides have been eluviated). Much of the descriptive terminology is also adopted from the USDA Comprehensive System, but

Table 14.2. Higher classes from the World Soil Map with some included classes from national classifications.

Fluvisols	(L. *fluvius*, river Fr. *sol*, soil) connotative of flood plains and alluvial deposits; incl. Alluvial soils (Aust.), Sols tropicaux récents, Fluvents (USA).
Gleysols	(Russian work for mucky soil mass) connotative of excess water; incl. Haplaquepts, Humaquepts (USA).
Regosols	(Gr. *rhegos*, blanket) implying a loose mantle overlying a hard rock core; incl. Skeletal soils (Aust.), Orthents (USA).
Lithosols	(Gr. *lithos*, stone) connotative of very shallow soils over rock.
Arenosols	(L. *arena*, sand) suggesting immature coarse-textured soils; incl. Psamments (USA).
Rendzinas	(Po. *rzedzic*, noise) connotative of a plough scraping through a stony limestone soil; incl. Rendolls (USA).
Rankers	(Aus. *Rank*, steep slope) hence shallow soils but on siliceous parent material.
Andosols	(Jap. *An*, dark and *Do*, soil) material rich in volcanic glass commonly having a dark surface horizon; incl. Andepts (USA).
Vertisols	(L. *verto*, turn) connotative of churning surface soil; incl. Black Earths (Aust.).
Solonchaks	(R. *sol*, salt) connotative of salinity; incl. Salorthids (USA).
Solonetz	Salt-affected; incl. Natrustalfs, Natrixeralfs (USA).
Yermosols	(L. *eremus*, desolate) hence wide open, dry spaces; incl. Desert loams (Aust.), Typic Aridisols (USA).
Xerosols	(Gr. *xerox*, dry) desert environments; incl. Mollic Aridisols (USA).
Kastanozems	(L. *castaneo*, chestnut; R. *zemlja*, earth) soils rich in organic matter and brown or chestnut in colour; incl. Ustolls (USA).
Chernozems	(R. *chern*, black) soils rich in organic matter and black in colour; incl. Haploborolls, Vermiborolls (USA).
Phaeozems	(Gr. *phaios*, dusky) soils rich in organic matter with a dark A horizon; incl. Hapludolls (USA), Degraded Chernozems (R.).
Greyzems	(English *grey*, the colour) soils rich in organic matter and grey in colour; incl. Argiborolls, Aquolls (USA).
Cambisols	(L. *cambiare*, to change) suggesting changes in colour, structure and consistence due to weathering; incl. Chocolate soils (Aust.), Eutrochrepts, Ustochrepts (USA).
Luvisols	(L. *luvi*, washed) connotative of lessivage; incl. many podzolic soils, Red Brown Earths (Aust.), Hapludalfs, Haplo-xeralfs (USA).
Podzoluvisols	Intermediate in development between *Luvisols* and *Podzols*; incl. Glossudalfs (USA).
Podzols	(R. *pod*, under; *zola*, ash) connotative of soils with strongly bleached subsurface horizons; incl. Orthods, Ferrods (USA)
Planosols	(L. *planus*, flat) related to the flat or depressed topography with poor drainage in which these soils form; incl. Soloths, Solods (Aust., R.), Albaqualfs (USA).
Acrisols	(L. *acris*, very acid) hence low base saturation; incl. Hapludults (USA); Red-yellow podzolics (Aust.).
Nitosols	(L. *nitidus*, shiny) connotative of bright shiny ped faces; incl. Krasnozems (Aust.), Tropudalfs and Paleudalfs (USA).
Ferralsols	(L. *ferrum*, iron and *aluminium*) soils high in sesquioxides; incl. Sols ferralitques, Oxisols (USA).
Histosols	(Gr. *histos*, tissue) soils rich in fresh or partly decomposed organic matter; incl. Moor peats (Aust.), Sols hydromorphes organiques, Histosols (USA).

Abbreviations used: L. Latin, Gr. Greek, Po. Polish, Aus. Austrian, Jap. Japanese, R. Russian, Fr. French, Aust. Australian, incl. including.

as many as possible of traditional Great Group and Great Soil Group names have been retained, as well as new names coined which do not suffer in translation nor have different meanings in different countries. The units are mapped as soil associations, designated by the dominant soil unit, with three *textural* classes (coarse, medium and fine) and three *slope* classes superimposed (level to gently undulating, rolling to hilly and steeply dissected to mountainous).

For the sake of logical presentation and generalization at a higher level, the soil units have been grouped "on the basis of generally accepted principles of soil formation" to form the 26 'World Classes' which are set out in Table 14.2. Although formulated on inferred pedogenic processes such as gleying, salinization and lessivage, which have a bearing on soil use, the World Classes have been compiled so broadly and the total variation within each class is so large that their value in the prediction of land use is limited. Their merit is that they constitute a gallery of conceptual classes which convey meaning to a worldwide clientele.

14.4 SUMMARY

Soil classification is necessary for the orderly presentation to soil users of the accumulated knowledge of soil behaviour. Classification involves grouping soils into classes according to the degree of similarity between individuals. It is made more difficult and the results contentious because there is no definable soil entity—the soil mantle varies continuously so that subdivision of the variation in one or more properties to create classes is usually arbitrary.

Soil survey starts with the assessment of the range of variability and description of the soil in the field. This is followed by the creation of a *list of classes* (the classification) which allows any soil in the area to be placed in one class, and ends with the presentation of results, usually as a *soil map*. Depending on the map scale, which may vary from 1:1,000,000 for reconnaisance surveys to 1:1,000 for intensive surveys of management problems, surveying proceeds on a grid (for large scales) or free survey pattern (for small scales). Interpretation of air photos is helpful at small scales in locating soil

boundaries which coincide with landscape features.

Classes are defined in terms of one or two soil properties relevant to a specific purpose or on many properties that may be relevant for various soil uses. *Diagnostic* or *key* properties, which are not necessarily relevant to usage but are easily assessed in the field, are often chosen to define classes on the assumption they are correlated with the more important properties. The classes recognized by the surveyor (*soil series*) are usually aggregated to form progressively less homogeneous classes which facilitate national and international exchange of soil knowledge. Because of the wide choice of criteria on which to separate classes at any one level and to establish the different levels or categories of generalization, many classifications have been proposed. The World Soil Classification is recommended here for teaching soil science: it comprises 106 soil units grouped into 26 World Classes on the basis of generally accepted concepts of soil formation.

REFERENCES

BALDWIN H., KELLOGG C.W. & THORP J. (1938) Soil classification in *Soils and man. Yearbook of Agriculture, Washington D.C.*
FAO-UNESCO (1974) *Soil map of the World 1 : 5,000,000 Volume 1, Legend*, Paris.
SOIL SURVEY STAFF USDA (1960) *Soil classification, a comprehensive system, 7th approximation.* Washington, D.C.

FURTHER READING

BECKETT P.H.T. (1976) Soil survey. *Agricultural Progress* **51**, 33–49.
BECKETT P.H.T. & WEBSTER R. (1971) Soil variability—a review. *Soils and Fertilizers* **34**, 1–15.
MULCAHY M.J. & HUMPHRIES A.W. (1967) Soil classification, soil surveys and land use. *Soils and Fertilizers* **30**, 1–8.
VAN DER EYK J.J., MACVICAR C.N., & DE VILLIERS J.M. (1969) Soils of the Tugela Basin—a study in subtropical Africa. *Natal Town and Regional Planning Reports, Volume* **15**, Pietermaritzburg.
WEBSTER R. (1977) *Quantitative and Numerical Methods in Soil Classification and Survey*. Oxford University Press, Oxford.
YOUNG A. (1976) *Tropical Soils and Soil Survey*. Cambridge University Press, Cambridge.

Index

HID   EEK!

For River & Sky – Wherever you roam, there will always be hugs and toast when you come home – RB

For my Gran and Grandad, Sheila and Stuart – NK

PUFFIN BOOKS

UK | USA | Canada | Ireland | Australia
India | New Zealand | South Africa

Puffin Books is part of the Penguin Random House group of companies
whose addresses can be found at global.penguinrandomhouse.com.

www.penguin.co.uk   www.puffin.co.uk   www.ladybird.co.uk

Penguin
Random House
UK

First published 2022
001

Printed in China

The authorized representative in the EEA is Penguin Random House Ireland,
Morrison Chambers, 32 Nassau Street, Dublin D02 YH68

A CIP catalogue record for this book is available from the British Library

ISBN: 978–0–241–48696–2

All correspondence to:
Puffin Books
Penguin Random House Children's
One Embassy Gardens, 8 Viaduct Gardens, London SW11 7BW

FSC
www.fsc.org

MIX
Paper from
responsible sources
FSC® C018179